CHEMOKINE RESEARCH TRENDS

CHEMOKINE RESEARCH TRENDS

LEAH R. GRINWALD
EDITOR

Nova Science Publishers, Inc.
New York

LIBRARY OF CONGRESS CATALOGING-IN-PUBLICATION DATA
Chemokine research trends / Leah R. Grinwald, editor.
p. cm.
Includes bibliographical references and index.
ISBN-13: 978-1-60021-588-9 (hardcover)
ISBN-10: 1-60021-588-2 (hardcover)
1. Chemokines- -Immunology. I. Grinwald, Leah R.
[DNLM: 1. Chemokines- -immunology. QW 568 C5158 2007]
QR185.8.C45C472 2007
616.07'9- -dc22 2007007501

Published by Nova Science Publishers, Inc. ✦ New York

CONTENTS

PREFACE

Chemokines are the cytokines that may activate or chemoattract leukocytes. Each chemokine contains 65 ~ 120 amino acids, with molecular weight of 8 ~ 10 kD. Their receptors belong to G-protein-coupled receptors. Inflammatory chemokines are released from a wide variety of cells in response to bacterial infection, viruses and agents that cause physical damage such as silica or the urate crystals that occur in gout. They function mainly as chemoattractants for leukocytes, recruiting monocytes, neutrophils and other effector cells from the blood to sites of infection or damage. They can be released by many different cell types and serve to guide cells involved in innate immunity and also the lymphocytes of the adaptive immune system. The cells that are attracted by chemokines follow a signal of increasing chemokine concentration to the site of infection or tissue injury. Some chemokines also have roles in the development of lymphocytes, migration and angiogenesis (the growth of new blood vessels).Since the entry of HIV into host cells requires chemokine receptors, their antagonists are being developed to treat AIDS.

Expert Commentary- The interferon γ-inducible, non-ELR CXC chemokines have long been considered as factors with an anti-tumorigenic potential, acting on the tumor microenvironment *via* their "immuno-angiostatic" properties. However, recent findings suggest that these chemokines have an opposite effect, by acting directly on the tumor cells and promoting their pro-tumorigenic properties. In this commentary, the contradicting findings on the activities of these chemokines in malignancy are discussed, and their potential implications for therapy are addressed.

Chapter 1 - Although tuberculosis is a leading infectious disease affecting millions of people worldwide, no effective vaccine exists. Therefore, it is important that prospective vaccines engender maximal immunity. Genetic vaccines encoding adjuvants molecules represent one approach to optimize systemic immunity. Promising candidates include various inflammatory chemokines that might be used to enhance the immune response to a level of sufficient for protection, since chemokines are crucial during mycobacterial infections for a timely recruitment of specific leukocyte subpopulations to sites of tissue damage. Notable chemokines that may offer hope in such efforts include CCL21, RANTES, CCL19, and CXCL8. Even though CXCL8 is known to be important both in enhancing local antimycobacterial activity and in driving recruitment of leukocyte subsets during inflammatory processes, little information is known about the relationship between transcriptional upregulation of the mycobacteria-induced CXC-chemokine secretion and TLR signaling. In this chapter, the authors review current progress in the understanding of chemokine-driven mycobacteria infection. In addition, they discuss the potential involvement of Toll-like receptors (TLRs) and their signaling pathways in regulating chemokine secretion in mycobacterial infection. The topics covered in this chapter are ripe for future research.

Chapter 2 - Type I interferon (IFN) subtypes initiate the transcription of IFN-inducible gene products by binding to a shared receptor that is broadly expressed on the surface of most cells. Although the receptor for each subtype is communal, the specific downstream events initiated by each subtype are distinct. The purpose of this study was to compare and contrast the induction or augmentation of chemokine production by different classes of type I IFNs in HSV-1 infected and uninfected trigeminal ganglion cell cultures. Overall, the following observations were made: 1) CCL5 was induced by IFN-α1, -α4and-α9; 2) CXCL10 production was induced by IFN-α1, -α4, -α5, and -β;3) CCL2 production was augmented by IFN-α1, -α4, -α9, and -β. In all instances, HSV-1 infection was associated with a reduction in IFN-induced chemokine production. Abridged chemokine production was associated with suppression of cytokine signaling-3 expression rather than apoptosis. Collectively, the results suggest HSV-1 suppresses chemokine production by TG cells identifying another means by which the virus circumvents the host's immune response.

Chapter 3 - Worldwide, co-infection by Mycobacterium tuberculosis is common among HIV-1-infected subjects. These two pathogens influence the prevalence, morbidity, and mortality caused by one another. Tuberculosis (TB) occurs earlier during HIV-1 disease (CD4 cell count over 200/μl) in two thirds of dually infected subjects, and is associated with accelerated rates of HIV-1 disease progression. Higher viral loads, increased HIV-1 diversity, and changes in cytokine and chemokine levels at sites of M. tuberculosis infection (pleural or pulmonary TB) have been well-documented in dually infected subjects and likely all contribute to the negative impact of TB on HIV disease. Increased expression of CCR5 in combination with relatively low levels of HIV-1-inhibitory β chemokines is conducive to HIV-infection of lymphocytes and macrophages that have recently migrated to M. tuberculosis infected organs. Together with accentuated HIV-1 replication, which is guaranteed by excesses in TNF-α and MCP-1, a state of saturation of infection by CCR5-tropic HIV in *situ* results. Due to excessive cellular activation, however, preferential deletion of these HIV-infected cells may occur. On the other hand, increased expression of CXCR4 by macrophages supported by excess TGF-β1 in situ supports enhanced infection by CXCR-4 tropic HIV-1. This may allow expansion of the more cytopathic viral strains in *situ*, and ultimately contribute to a predominance by CXCR-4 tropic HIV-1 infection. Increased levels of IFN-γ and TNF-α in *situ* guarantee cellular activation, which is necessary for successful infection through either chemokine receptor. This review addresses the role of chemokines and their receptors at sites of active tuberculosis in progression of HIV-1 infection in dually infected subjects.

Chapter 4 - Rheumatoid arthritis (RA) is a chronic, symmetric polyarticular joint disease characterized by synovial proliferation and the progressive destruction of cartilage and bone. We found that the expression of pulmonary and activation-regulated chemokine (PARC)/CCL18 and monocyte chemoattractant protein-4 (MCP-4)/CCL13 were significantly higher in cartilage from RA patients than from osteoarthritis (OA) patients or normal controls. Although the biological function of CCL18/PARC has not been elucidated, the authors found high expression in chondrocytes and synovial cells of RA patients, and determined that serum CCL18/PARC levels significantly correlated with rheumatoid factor. With regard to MCP-

4/CCL13, we found that IFN-γ significantly stimulated MCP-4/CCL13 production from human chondrocytes in a dose-dependent manner and that this production was significantly enhanced in combination with IL-1β or TNF-α. MCP-4/CCL13 induces the phosphorylation of ERK in fibroblast-like synoviocytes (FLS) and activates the proliferation of these cells. A specific inhibitor of ERK kinase, PD98059, completely inhibited the stimulatory effect of MCP-4/CCL13 on FLS proliferation. From these data they concluded that IFN-γ in combination with IL-1β or TNF-α activates the production of MCP-4/CCL13 from chondrocytes in RA joints, and secreted MCP-4/CCL13 enhances FLS proliferation by activating the ERK MAP kinase cascade resulting in RA pathogenesis.

Systemic lupus erythematosus (SLE) is an autoimmune disease characterized by widespread immunologic abnormalities and multiorgan involvement, including the skin, joints, kidney, serosal membranes, and peripheral and central nervous system (CNS). Neuropsychiatric systemic lupus erythematosus (NPSLE) is a major contributor to morbidity and mortality in patients with (SLE). These results showed that cerebral spinal fluid (CSF) levels of Th-1 type chemokines such as IP-10/CXCL10 and MCP-1/CCL2 were higher in NPSLE patients than in non-NPSLE patients. This suggests that CSF IP-10/CXCL10 and MCP-1/CCL2 reflect inflammatory activity in the brain, and may be used as a diagnostic tool as well as a gauge for monitoring therapeutic responses in patients with NPSLE.

Chapter 5 - Leukocytosis is a well-known hematological disorder in tobacco smokers. This malady includes neutrophilia, lymphocytosis, and additionally, under a certain analysis, monocytosis and eosinophilia. The cause of the leukocytosis, however, has yet to be clarified. One of the reasons may be due to the fact that it remains difficult to account for such a pattern of leukocytosis with the *in vivo* biologic action of single hematopoiteic cytokine. The authors have attempted to identify the cause of leukocytosis with respect to blood chemokine levels and the effect of nicotine, a major substance in tobacco smoke, on the chemokine production by white cells *in vitro*. The blood levels of IL-8, MCP-1, and RANTES were elevated in smokers. Nicotine promoted IL-8 and MCP-1 production by neutrophils as well as RANTES production by T lymphocytes. Activation of NF-κB was involved in the production of these chemokines. However, signals upstream NF-κB differed, i.e., neutrophil IL-8 production was mediated by ROIs, but ROIs were only partially involved in the RANTES production by T cells. Based on the examination for the smokers who smoke 20 cigarettes and more a day and have 10,000/μL leukocytes or more, we propose that the levels of IL-8 and RANTES in the blood could be adequate markers of heavy smoking with leukocytosis.

Chapter 6 - Given the rapid progress being made in the understanding of the migration of leukocytes to inflammatory sites, research aiming to target proinflammatory chemokines and their receptors in patients with rheumatoid arthritis (RA) and animal models of RA has been actively pursued in the last few years. Antagonists of a number of chemokines and chemokine receptors are in phase 1 and 2 trials for RA. In previous studies the expressions of chemokine receptors CCR1 to 9, CXCR1 to 5, CX_3CR1, and XCR1 were examined in mononuclear cells (MNCs) of paired synovial fluid and blood samples from RA patients by quantitative reverse transcription polymerase chain reaction. Highly elevated levels of CCR5 and XCR1 in synovial MNCs were consistently demonstrated in all patients. Moreover, levels of the corresponding ligands, MIP-1α and lymphotactin, were increased in synovial fluid. Therefore, in this study we examined the involvement of CCR5 and XCR1in the DBA/1 mouse model and MIP-1α in the Sprague-Dawley rat model of collagen-induced arthritis (CIA). Greater numbers of $CCR5^+$ T cells infiltrated more severe inflamed joints, with a concomitant increase in MIP-1α production. In addition, MNCs from more severe inflamed joints expressed higher levels of XCR1. Depletion of $CCR5^+$ cells significantly decreased type II collagen-induced T cell proliferation and MIP-1αproduction. Adenovirus-mediated gene therapy that suppressed the MIP-1αpromoter activity significantly reduced MIP-1αlevels in ankle joint extracts and suppressed CIA. Taken together, our results from animal experiments suggest a therapeutic potential of targeting MIP-1α and its receptor CCR5 in RA patients.

Chapter 7 - Neuropathic pain is the chronic pain state caused by pathological changes in the nervous system. It can occur secondarily to injury of the central nervous system but it occurs most commonly in association with injury to the peripheral nervous system. The pathogenesis of neuropathic pain involves neural-immune interactions that activate immune cells, glial cells and neurons. In this chapter, the authors summarize recent findings regarding the contribution of monocyte chemoattractant protein-1 (MCP-1), a CC family chemokine, in the etiology of this pathology. Constitutively present in the immune system, MCP-1 could be induced in damaged sensory neurons; its receptor, CCR2, is also present within the central nervous system. Under the powerful chemoattractant effect of MCP-1, monocytes/macrophages are recruited around the injured nerve and proinflammatory substances released by these immune cells increase the peripheral sensitization. In addition to its classical function in cell trafficking, MCP-1 could contribute to central sensitization by directly increase neuronal excitability or by modulation of spinal inhibitory GABAergic transmission. It is also suggested that MCP-1 could be an important molecule mediating neuron-glia interaction and trigger spinal glial activation. Binding of MCP-1 to CCR2 activates different signaling pathways upon the cellular environment and may define their physiopathological effect. These emerging observations provide new insights into the full spectrum of neuropathic pain and present continuing challenges for drug discovery.

Chapter 8 - Receptor activator of NF-kB ligand (RANKL) plays a crucial role in the molecular mechanisms of bone metabolism in bone-related diseases, such as osteoporosis and also in inflammatory conditions characterized by local bone remodelling including rheumatoid arthritis. In fact, RANKL is essential for differentiation of human monocytes into osteoclasts and also, it is a potent chemotactic factor for human monocytes with an efficacy similar to classic monocyte chemoattractants, such as CCL2 and CCL3. Nevertheless, the intracellular kinase pathways triggered by RANKL, CCL2 and CCL3 are still not well known and controversial. To clarify this issue, we investigated monocyte migration and kinase activation to the aforementioned ligands. Chemotaxis of human monocytes was performed in modified Boyden chamber with polycarbonate filter. Intracellular kinase activation was analyzed by western blot assay. These results show that:

1. RANKL, CCL2 and CCL3 induce monocyte migration via a common pathway involving Src kinases, as suggested by the capacity of PP1 Analog (inhibitor of Src kinase) to inhibit monocyte migration to the three cytokines. On the contrary, although activated by each chemoattractant, ERK 1/2 activation is unnecessary for monocyte migration, as suggested by the lack of activity of PD 98059, the inhibitor of MEK 1/2 (a kinase activating ERK 1/2).

2. RANKL-induced monocyte locomotion is strictly related to PI3-K/Akt and p38 MAPK activity, but independent from JNK 1/2 activation. Indeed, LY 294002 (inhibitor of PI3-K), Akt inhibitor and SB 203580 (inhibitor of p38 MAPK) inhibited monocyte migration to

RANKL, whereas SP600125 (inhibitor of JNK1/2) was ineffective. Accordingly, RANKL activated Akt and p38 MAPK, but not JNK 1/2..

3. CCL2 and CCL3 monocyte stimulation do not induce Akt, p38 MAPK and JNK 1/2 activation. Consistently, chemokine-driven monocyte migration is unaffected by LY 294002, Akt inhibitor, SB 203580, an SP600125.

These results provide evidence that RANKL- and CC chemokines-induced monocyte migration share a common pathway of signal transduction mediated by the activation of Src kinases. Nevertheless, RANKL delivers its locomotory signal via a transduction pathway downstream Src kinases, which is distinct from that of CCL2 and CCL3, PI3-K/Akt and p38 MAPK being selectively activated only by RANKL. The intracellular pathway(s) distal to Src and involved into CC chemokine-dependent migration remains to be identified. Thus, PI3-K/Akt and p38 MAPK may be proposed as possible targets for the pharmacologic inhibition of RANKL-mediated monocyte activation, without impairing monocyte locomotory response to classical chemokines generally involved in anti-infective responses.

Chapter 9 - Malignant melanoma may arise de novo or from potential precursor lesions, such as atypical dysplastic nevi or congenital nevi. The ability of tumour cells to avoid immune surveillance represents a critical step in tumour progression. It has been suggested that the expression of chemokines and chemokine receptors by melanoma cells may be involved in tumour immune escape. Recently ligation of tumor-associated CCR10 with CCL27 has been proposed as a mechanism that promotes neoplastic cell survival and reduce immune response in some animal models. The aim of this study was to investigate, by immunohistochemistry, the relation between T immunoreactive cells and the expression of CCR10 and its ligand CCL27 in human melanocytic lesions.

CCL27 and CCR10 expression in melanocytes and the density of CD4+, CD8+ and CD3+ lymphocytes have been assessed in 59 cutaneous melanocytic lesions: 40 malignant melanomas, 10 Spitz nevi, and 9 benign melanocytic nevi.

In malignant melanomas T lymphocyte density was significantly different in the three groups of cases selected by different thickness, showing tendency to decrease from thin melanomas to intermediate and thick ones (Kruskal-Wallis test, $P<0.0005$). CCR10 expression was found both in benign and malignant lesions, being moderate and focal in 15 cases and strong and diffuse in 37. The percentage of expression did not differ between nevi, Spitz nevi and melanomas (Kruskal-Wallis test, $P=0.17$). However, in melanomas with metastatic sentinel lymph node CCR10 mean expression was significantly higher than in cases with negative lymph node (84.1% vs 58.1% in metastatic cases; unpaired t test, $P=0.0281$). Moreover in melanomas CCR10 shows a tendency to be directly correlated with the Breslow depth ($r=0.343$; $P=0.0298$) and inversely with T lymphocyte density ($r=-0.363$; $P=0.0231$).

In all benign lesions, CCL27 was negative or faintly expressed by single melanocytes, while in melanomas it was negative in 22 cases, weakly expressed in 16 and strongly and diffusely expressed in 2. When CCR10 and CCL27 expression were evaluated together, CCR10-/CCL27- melanomas tended to have a higher mean density of CD3+ and CD8+ lymphocytes.

The results confirm that also in human melanomas CCR10 and CCL27 may act to increase the ability of neoplastic cells to grow, invade tissue, disseminate to lymph nodes and to escape the host immune response.

Chapter 10 - The chemokine SDF1(CXCL12) and its receptor CXCR4 play pivotal roles in survival and metastasis of cancer cells. Recent studies have identified functional cross talk between the SDF1/CXCR4 pathway and several oncogenes that regulate migration, invasion, survival, trafficking and metastasis of transformed cells.

The receptor tyrosine kinase, HER2, which is frequently over-expressed in breast cancer cells and associated with an adverse prognosis, up-regulates CXCR4 expression and migration and invasiveness of breast cancer cells in response to SDF1. Moreover, CXCR4 is required for HER2-mediated metastasis to lung in vivo. Conversely, SDF1 transactivates HER2 in breast cancer cells. These results implicate the interaction of HER2 and SDF1/CXCR4 signaling pathways in the regulation of breast cancer progression and metastasis.

Similar to metastasis by solid tumor cells, leukemia cells leave the marrow and accumulate in the spleen and blood and often infiltrate extramedullary sites, suggesting that signaling events that regulate migration, homing and mobilization may be deregulated by the genes associated with hematological malignancies. Elevated CXCR4 expression is associated with Internal Tandem Duplication of the tyrosine kinase Flt3 (ITD-Flt3) and is also associated with poor prognosis in patients with AML. SDF1/CXCR4 and ITD-Flt3 functionally interact to enhance hematopoietic cell survival and migration. Furthermore, ITD-Flt3 decreases adhesion of hematopoietic cells to VCAM-1 and facilitates their accumulation in the spleen, suggesting that ITD-Flt3 modulates trafficking of transformed hematopoietic cells. In contrast to ITD-Flt3, the oncogene BCR/ABL, which is found in patients with Chronic Myeloid Leukemia (CML), down regulates CXCR4 expression and migration to SDF1, which may contribute to defects in retention and homing of CML cells.

Understanding the mechanisms by which oncoproteins regulate SDF1/CXCR4 signaling and function will facilitate development of novel strategies to block metastasis and eliminate development of resistance to apoptosis characteristic of cancer cells and leukemia cells. This chapter will summarize the recent evidence on signaling cross talk between oncoproteins and SDF/CXCR4 pathways that may contribute to the dissemination, metastasis and enhanced survival of cancer cells and leukemia cells.

Chapter 11 - Chemokines are a superfamily of chemotactic cytokines. Recently, platelets have been shown to contain members of the chemokine family, both of CXC class, such as interleukin (IL)-8 and epithelial-neutrophil activating peptide (ENA)-78, and of the C-C class, such as regulated on activation normally T-cell expressed and secreted (RANTES). RANTES is a potent chemoattractant of memory T lymphocytes, monocytes, eosinophils and basophils. And ENA-78 is chemokine known to be chemoattractant for neutrophils. The authors measured and compared levels of platelet activation markers, platelet-derived chemokines, and soluble factors in patients with various diseases and clinical settings. RANTES differed in the healthy control and hypertensive patients, particularly with type 2 diabetes. In addition, this chemokine was higher in the patients with acute myocardial infarction than in patients with stable pectoris angina. ENA-78 concentration was found in Helicobacter pylori-positive patients with immune thrombocytopenic purpura. Furthermore, ENA-78 and RANTES exhibited a significant elevation after stem cell transplantation. Thus, platelet-derived chemokines demonstrate the role of platelets as inflammatory cells in various clinical settings.

Chapter 12 - The recently discovered chemokine/scavenger receptor CXCL16 is an interesting chemokine molecule, which is emerging as a highly significant player in atherosclerosis, rheumatoid arthritis (RA), neuroinflammatory diseases and other conditions.

CXCL16 was first identified as a scavenger receptor for oxidized LDL (oxLDL). It was shown to play a role in the internalization of oxLDL particles, leading to subsequent foam cell formation, which is a critical step in the development of atherosclerotic plaques. It is a transmembrane molecule, which is expressed in dendritic cells, macrophages and aortic smooth muscle cells and isaos cleaved by the metalloprotease ADAM-10 to release a soluble form that has chemotactic activity and is the only known ligand of the chemokine receptor CXCR6 which is expressed on a subset of T cells and NKT cells. CXCL16 may therefore play an important role in enhancing effector functions of T-cells.

CXCL16 has been detected in human and murine atherosclerotic lesions, and elevated levels of CXCL16 in cerebrospinal fluid suggest local production of CXCL16 in the central nervous system. It is also expressed in the synovium of RA patients. This chapter identifies important roles for CXCL16 in T-cell triggering and activation of effector functions, and describes research on understanding its significance in atherosclerotic vascular disease, immune responses in the central nervous system and the infiltration of leukocytes into synovial tissue and fluid in patients with the chronic inflammatory disease RA.

Chapter 13 - Chorioamnionitis, a major cause of pre-term labor, is generally diagnosed by histological examination of placenta tissues obtained from patients after delivery. Here the authors focus on use of amniotic fluid, which can be readily obtained in the clinical field during pregnancy, for diagnostic purposes. Culture is the gold standard for detection of infection, but bacteria often fail to be detected by routine culture in amniotic fluid obtained from patients complicated by pre-term labor, despite increase in infectious cytokines, such as interleukin-6 and interleukin-8. The authors have established detection of Escherichia coli (E coli), one major cause of chorioamnionitis, by a modified polymerase chain reaction technique.

There is considerable evidence that NO inhibits uterine contractility as well as cervical ripening during pregnancy, but in our study, the NO concentration was found to be increased in amniotic fluid obtained from patients complicated by chorioamnionitis. These results indicate that effects of NO may be attenuated for at least two reasons: the action of NO itself is down-regulated and the NO is rapidly degraded in the amniotic fluid.

Chorioamnionitis is a major cause of pre-term labor. As an alternative to histological examination of placenta tissues obtained from the patients after delivery it has been proposed that amniotic fluid be used for testing. For example, evaluation of inflammatory parameters, such as the presence of white blood cells [1.2.3], interleukin-6 [1.3-6], interleukin-8 [1.6.7] or leukocyte elastase [1.4.8.9] has been applied for diagnosis of intrauterine infection.

In the present report, the authors introduce new concepts for amniotic fluid obtained from patients complicated by chorioamnionitis.

Chapter 14 - Endothelial cell (EC) dysfunction has been proposed to be an initial event that leads to the development of atherosclerosis. There is a growing body of evidence that atherosclerosis is an inflammatory disease, and ECs are target for chemokines and growth factors released during inflammation. Recently, nifedipine, one of the most widely used dihydropyridine-based calcium antagonists (DHPs) for the treatment of patients with hypertension, has been shown to inhibit vascular inflammation and subsequently improve endothelial dysfunction in many cardiovascular diseases, thus slowing the development and progression of atherosclerosis. However, the molecular mechanisms underlying this are not fully understood, because ECs do not possess voltage-operated L-type calcium channels. In this review,the authors discuss molecular mechanisms by which nifedipine could exert anti-inflammatory effects on ECs.

In: Chemokine Research Trends
Editor: Leah R. Grinwald, pp. 1-5

ISBN: 978-60021-588-9

Expert Commentary

THE CHEMOKINE RECEPTOR CXCR3 AND ITS LIGANDS IN MALIGNANCY: DO THEY ACT AS DOUBLE-EDGED SWORDS?

Adit Ben-Baruch
Dept. Cell Research and Immunology, George S. Wise Faculty of Life Sciences,
Tel Aviv University, Tel Aviv 69978, Israel

ABSTRACT

The interferon γ-inducible, non-ELR CXC chemokines have long been considered as factors with an anti-tumorigenic potential, acting on the tumor microenvironment *via* their "immuno-angiostatic" properties. However, recent findings suggest that these chemokines have an opposite effect, by acting directly on the tumor cells and promoting their pro-tumorigenic properties. In this commentary, the contradicting findings on the activities of these chemokines in malignancy are discussed, and their potential implications for therapy are addressed.

COMMENTARY

The chemokine superfamily includes a large number of members, with differing properties and chemotactic activities on defined leukocyte subsets. In addition to their well-characterized roles in the regulation of inflammation and immune homeostasis, many members of this superfamily were recently suggested to play a role in malignancy (1-4). In this context, chemokine activities are diverse, either potentiating or inhibiting tumor development and/or progression, depending on the chemokine, the tumor type, the cell system used, as well as other factors at the tumor microenvironment (1-4).

In the framework of malignancy, much emphasis was given to interferon γ (IFNγ)-inducible, non-ELR CXC chemokines: CXCL9, CXCL10 and CXCL11. These chemokines share the CXCR3 receptor, for which three different variants have been thus far identified in human: CXCR3-A, CXCR3-B and CXCR3-alt (5, 6). The above-mentioned CXCR3 ligands were shown to be chemotactic for a variety of leukocyte subsets, most effectively for Th1 cells (7-9). This characteristic of the three CXCR3 ligands has provided the first clue to their potential involvement in malignancy, more specifically by acting as Th1 chemoattractants, thus potentially giving rise to tumor inhibition.

In recent years, numerous investigations of the roles played by the three CXCR3 ligands in malignancy have been gathered into the concept of "immuno-angiostasis" (7, 10). This term reflects the ability of IFNγ-inducible CXCR3 ligands to exert concomitantly two tumor-inhibiting properties: [1] Attracting to tumor sites CXCR3-expressing mononuclear cells with anti-tumor activities; [2] Attenuating tumor-related angiogenesis.

Many lines of evidence support the "immuno-angiostasis" nature of the three IFNγ-inducible CXCR3 ligands. The ability of these chemokines to promote anti-tumor cell-mediated immunity, and at the same time to inhibit angiogenesis, results in suppression of tumor growth and metastasis, as was clearly evidenced in models of non-small cell lung cancer (NSCLC) and renal cell carcinoma (RCC) (7, 10, 11). Combined with findings in other tumor systems, the different studies have resulted in the categorization of the IFNγ-inducible non-ELR CXC chemokines as possible inhibitors of tumor formation and metastasis.

The immuno-angiostatic activities of the CXCR3 ligands actually reflect their ability to act on cells of the tumor microenvironment, be it leukocytes of different types or endothelial cells. However, recent findings indicate that CXCR3 expression is not limited to host cells at the tumor cell milieu, but rather that the receptor is expressed also by tumor cells. CXCR3 expression was detected on cancer cells of hematopoietic origin, such as multiple myeloma (MM), acute lymphoblastic leukemia (ALL) and B-cell chronic lymphocytic leukemia (B-CLL) (12-14). In addition, many cancer cell lines derived of solid tumors express CXCR3 receptors, including breast carcinoma, RCC, neuroblastoma, melanoma cells, colorectal carcinoma (CRC) cells and more. Recent studies also indicate that CXCR3 expression is not limited to tumor cell lines, as it is also detected in clinical samples derived from breast tumors, RCC, melanoma and CRC (15-24).

As CXCR3 is a receptor for chemokines, being active primarily in the immune context, its expression by hematopoietic tumor cells may manifest the cellular origin of these cells, belonging to the hematopoietic lineage. However, the expression of CXCR3 by cancer cells that originated from solid tumors is puzzling: Why do such tumor cells express receptors for chemokines that are tumor-limiting in their nature? Is it possible that the tumor cells have undergone selection that enables them to take advantage of tumor-supporting constituents that are present at the tumor microenvironment? Could it be that the tumor cells exploit these chemokines for their own benefit?

Emerging findings suggest that this may indeed be the case. In several tumor systems derived from solid malignancies, IFNγ-inducible CXCR3 ligands were found to induce intracellular signaling and to promote cellular properties that may improve the tumorigenic potential of the cells. With regard to signaling, the IFNγ-inducible non-ELR CXC chemokine/s were found to regulate Erk, p38 and/or PI3K activation in thyroid cancer cells, melanoma, neuroblastoma cells and CRC cells, to induce calcium mobilization in breast tumor cells and to activate Rho A and Rac 1 in melanoma cells (16, 19, 20, 23, 25). When tumor-promoting properties are concerned, several such functions were found to be elicited by these chemokines in tumor cells (e.g. thyroid, breast, melanoma, CRC), including induction of matrix metalloproteinase expression, of actin polymerization, of migration, and of invasion (16, 20, 21, 23, 25). In CRC, CXCL10 was found to act specifically on metastatic human CRC cells, inducing their migration, with no such activity on primary CRC cells. The migratory phenotype of the metastatic cells was accompanied by CXCL10-promoted adhesion, thus suggesting that CXCR3-CXCL10 interactions may support metastatic processes at remote sites in which CXCL10 is expressed (such as liver) (23). Furthermore, in several of the tumor cell systems CXCR3 ligands were found to promote proliferation (21, 25), thus serving as important potential tumor growth stimulators. Similar findings were made also in hematopoietic tumors, e.g. MM cells were found to respond to all three IFNγ-inducible CXCR3 ligands by migration, and by elevated production of matrix metalloproteinases (12).

These *in vitro* findings suggest that CXCR3 ligands that are present at the tumor milieu, may act *in situ* as tumor-promoting entities *via* their activity on the tumor cells. They actually suggest that these chemokines may have opposing effects on cells at the tumor microenvironment *vs* the tumor cells themselves: while acting as immuno-angiostatic factors on the tumor microenvironment, they act directly on the tumor cells to promote their pro-malignancy functions. The question is then which of the two opposing effects of these chemokines dominates?

The answer to this question is far from being at hand, since a large variety of factors may dictate the effects of CXCR3 ligands on cancerous processes, including not only the type of tumor addressed but also the type of CXCR3 variant/s expressed by the tumor cells, the level and composition of the IFNγ-inducible CXCR3 chemokines which are expressed at the tumor site, and their ability to exert immuno-angiostatic or tumor-promoting functions *in situ*. Whereas earlier studies have provided much support to the major roles played by these chemokines in protection from malignancy, recent findings indicate that the opposite may occur in particular occasions. More specifically, studies performed in animal model systems and on clinical samples from cancer patients provide preliminary evidence to the tumor-promoting functions of the IFNγ-inducible CXCR3 chemokines.

Prominent pro-malignancy roles for the axis of CXCR3-CXCR3 ligands were found in animal model systems of murine melanoma and of breast carcinoma, showing that inhibition of CXCR3 expression/activities has given rise to significant reduction in metastasis (16, 21). These observations are in line with previous findings on the ability of these chemokines to directly affect melanoma and breast carcinoma cells, and to induce tumor-supporting functions in these cells (as discussed above). Taking the enigma one step further, tumor-supporting roles for CXCR3 and its ligands were suggested in cancer patients. For example, in cutaneous malignant melanoma CXCR3 immunoreactivity was correlated with clinicopathologic prognostic factors (22). Similar roles for IFNγ-inducible chemokines and for CXCR3 were also implicated in CRC and RCC (18, 24). In the latter, up-regulation of CXCL11 and CXCL9, as well as of CXCR3 was detected in RCC tumors as compared to normal kidney tissues (18).

Actually, RCC provides a good example for the complexity of the issue, since in contrast to the above-mentioned study (18), a recent research has shown that CXCL9 has an independent favorable prognostic value in this disease (26). Such controversy, as evidenced in RCC, indicates that it is difficult to predict what will be the exact role of CXCR3 ligands in malignancy, even in one specific malignant disease. The existence of three different CXCR3 variants in human, but not in mouse, raises additional concerns: it is possible that models using murine tumor cells are not appropriate when we wish to learn more on human cancers, because different human tumor cells may express various combinations of the three CXCR3 variants. The situation may be further complicated by the ability of each of the variants to regulate differently specific cell functions that are related to malignancy, therefore further complicating our current view of the involvement of CXCR3 and its ligands in human malignancies.

Indeed, this possibility is supported by a recent study of human breast carcinoma cells, in which opposing functions were attributed to CXCR3-B as compared to another CXCR3 variant, probably CXCR3-A (17). Similarly, observations on the ability of CXCR3 ligands to promote tumor cell proliferation in some human tumor cell systems (21, 25), but to inhibit it in others (19, 27), may also reflect CXCR3 variant heterogeneity in different tumor cells.

Overall, the present information available on the roles of CXCR3 and its ligands in malignancy stresses the possibility that their effects on tumorigenesis and metastasis are much more diverse than earlier hypothesized. The IFNγ-inducible CXCR3 chemokines may have opposing effects on the tumor cells as compared to host cells at their proximity, at the tumor microenvironment. The ability of these chemokines to directly act on the tumor cells and to promote their tumor-supporting properties may antagonize the immuno-angiostatic effects they have. In view of the recent findings on their potential deleterious effects in malignancy, earlier hopes of using these chemokines as therapeutic means for inhibiting cancerous processes should be re-evaluated. Also, emphasis should be given to thorough research on CXCR3 and its ligands in a large variety of tumor systems, and to the elucidation of the implications of their expression in human malignancies.

References

[1] Balkwill F. Cancer and the chemokine network. *Nat Rev Cancer* 2004;4:540-50.

[2] Zlotnik A. Chemokines in neoplastic progression. *Semin Cancer Biol* 2004;14:181-5.

[3] Mantovani A, Allavena P, Sozzani S, Vecchi A, Locati M, Sica A. Chemokines in the recruitment and shaping of the leukocyte infiltrate of tumors. *Semin Cancer Biol* 2004;14:155-60.

[4] Ben-Baruch A. The multifaceted roles of chemokines in malignancy. *Cancer and metastasis reviews* In Press.

[5] Lasagni L, Francalanci M, Annunziato F, *et al.* An alternatively spliced variant of CXCR3 mediates the inhibition of endothelial cell growth induced by IP-10, Mig, and I-TAC, and acts as functional receptor for platelet factor 4. *J Exp Med* 2003;197:1537-49.

[6] Ehlert JE, Addison CA, Burdick MD, Kunkel SL, Strieter RM. Identification and partial characterization of a variant of human CXCR3 generated by posttranscriptional exon skipping. *J Immunol* 2004;173:6234-40.

[7] Strieter RM, Belperio JA, Burdick MD, Sharma S, Dubinett SM, Keane MP. CXC chemokines: angiogenesis, immunoangiostasis, and metastases in lung cancer. *Ann N Y Acad Sci* 2004;1028:351-60.

[8] Liu L, Callahan MK, Huang D, Ransohoff RM. Chemokine receptor CXCR3: an unexpected enigma. *Curr Top Dev Biol* 2005;68:149-81.

[9] Neville LF, Mathiak G, Bagasra O. The immunobiology of interferon-gamma inducible protein 10 kD (IP-10): a novel, pleiotropic member of the C-X-C chemokine superfamily. *Cytokine Growth Factor Rev* 1997;8:207-19.

[10] Strieter RM, Burdick MD, Mestas J, Gomperts B, Keane MP, Belperio JA. Cancer CXC chemokine networks and tumour angiogenesis. *Eur J Cancer* 2006;42:768-78.

[11] Pan J, Burdick MD, Belperio JA, *et al.* CXCR3/CXCR3 ligand biological axis impairs RENCA tumor growth by a mechanism of immunoangiostasis. *J Immunol* 2006;176:1456-64.

[12] Pellegrino A, Antonaci F, Russo F, *et al.* CXCR3-binding chemokines in multiple myeloma. *Cancer Lett* 2004;207:221-7.

[13] Wu S, Gessner R, Taube T, *et al.* Chemokine IL-8 and chemokine receptor CXCR3 and CXCR4 gene expression in childhood acute lymphoblastic leukemia at first relapse. *J Pediatr Hematol Oncol* 2006;28:216-20.

[14] Jones D, Benjamin RJ, Shahsafaei A, Dorfman DM. The chemokine receptor CXCR3 is expressed in a subset of B-cell lymphomas and is a marker of B-cell chronic lymphocytic leukemia. *Blood* 2000;95:627-32.

[15] Goldberg-Bittman E, Neumark, E., Sagi-Assif, O., Azenshtein, E., Meshel, T., Witz, I. P., Ben-Baruch, A. The expression of the chemokine receptor CXCR3 and its ligand, CXCL10, in human breast adenocarcinoma cell lines. *Immunol Lett* 2004;92:171-8.

[16] Walser TC, Rifat S, Ma X, *et al.* Antagonism of CXCR3 inhibits lung metastasis in a murine model of metastatic breast cancer. *Cancer Res* 2006;66:7701-7.

[17] Datta D, Flaxenburg JA, Laxmanan S, *et al.* Ras-induced Modulation of CXCL10 and Its Receptor Splice Variant CXCR3-B in MDA-MB-435 and MCF-7 Cells: Relevance for the Development of Human Breast Cancer. *Cancer Res* 2006;66:9509-18.

[18] Suyama T, Furuya M, Nishiyama M, *et al.* Up-regulation of the interferon gamma (IFN-gamma)-inducible chemokines IFN-inducible T-cell alpha chemoattractant and monokine induced by IFN-gamma and of their receptor CXC receptor 3 in human renal cell carcinoma. *Cancer* 2005;103:258-67.

[19] Goldberg-Bittman L, Sagi-Assif O, Meshel T, *et al.* Cellular characteristics of neuroblastoma cells: regulation by the ELR--CXC chemokine CXCL10 and expression of a CXCR3-like receptor. *Cytokine* 2005;29:105-17.

[20] Robledo MM, Bartolome RA, Longo N, *et al.* Expression of functional chemokine receptors CXCR3 and CXCR4 on human melanoma cells. *J Biol Chem* 2001;276:45098-105.

[21] Kawada K, Sonoshita M, Sakashita H, *et al.* Pivotal role of CXCR3 in melanoma cell metastasis to lymph nodes. *Cancer Res* 2004;64:4010-7.

[22] Monteagudo C, Martin JM, Jorda E, Llombart-Bosch A. CXCR3 chemokine receptor immunoreactivity in primary cutaneous malignant melanoma: correlation with clinicopathologic prognostic factors. *J Clin Pathol* 2006.

[23] Zipin-Roitman A, Meshel T, Sagi-Assif O, *et al.* CXCL10 promotes invasion-related properties in human colorectal carcinoma cells. *Cancer Res* 2007;67:3396-405.

[24] Kawada K, Hosogi H, Sonoshita M, *et al.* Chemokine receptor CXCR3 promotes colon cancer metastasis to lymph nodes. *Oncogene* 2007.

[25] Melillo RM, Castellone MD, Guarino V, *et al.* The RET/PTC-RAS-BRAF linear signaling cascade mediates the motile and mitogenic phenotype of thyroid cancer cells. *J Clin Invest* 2005;115:1068-81.

[26] Kondo T, Nakazawa H, Ito F, *et al.* Favorable prognosis of renal cell carcinoma with increased expression of chemokines associated with a Th1-type immune response. *Cancer Sci* 2006;97:780-6.

[27] Nagpal ML, Davis J, Lin T. Overexpression of CXCL10 in human prostate LNCaP cells activates its receptor (CXCR3) expression and inhibits cell proliferation. *Biochim Biophys Acta* 2006;1762:811-8.

In: Chemokine Research Trends
Editor: Leah R. Grinwald, pp. 7-28
ISBN: 978-60021-588-9

Chapter 1

CHEMOKINES AND MYCOBACTERIAL INFECTION: CRITICAL ROLES OF THE TLRS, AND THEIR SIGNALING PATHWAYS

Patricia Méndez-Samperio**, Laura Alba and Laura Belmont
Departamento de Inmunología, Escuela Nacional de Ciencias Biológicas, IPN.Carpio y Plan de Ayala, México, D.F. 11340 México.

ABSTRACT

Although tuberculosis is a leading infectious disease affecting millions of people worldwide, no effective vaccine exists. Therefore, it is important that prospective vaccines engender maximal immunity. Genetic vaccines encoding adjuvants molecules represent one approach to optimize systemic immunity. Promising candidates include various inflammatory chemokines that might be used to enhance the immune response to a level of sufficient for protection, since chemokines are crucial during mycobacterial infections for a timely recruitment of specific leukocyte subpopulations to sites of tissue damage. Notable chemokines that may offer hope in such efforts include CCL21, RANTES, CCL19, and CXCL8. Even though CXCL8 is known to be important both in enhancing local antimycobacterial activity and in driving recruitment of leukocyte subsets during inflammatory processes, little information is known about the relationship between transcriptional upregulation of the mycobacteria-induced CXC-chemokine secretion and TLR signaling. In this chapter, we review current progress in the understanding of chemokine-driven mycobacteria infection. In addition, we discuss the potential involvement of Toll-like receptors (TLRs) and their signaling pathways in regulating chemokine secretion in mycobacterial infection. The topics covered in this chapter underscore ripe for future research.

Key Words: Chemokines, cell signaling, infection, innate immunity

* E-mail address: pmendezs@bios.encb.ipn.mx, Tel: 5 729 60 00 ext 62499, Fax: (5) 396 35 03. Correspondence address: Dr. Patricia Méndez-Samperio, Departamento de Inmunología, Escuela Nacional de Ciencias Biológicas, IPN. Carpio y Plan de Ayala. México, D.F. 11340 México

INTRODUCTION

In the last several years, there has been substantial progress in the fields of chemokine research, and there is growing evidence that chemokines play an important role in the organization of the immune system [1, 2]. The chemokines are small secreted basic proteins (6-14 kDa) that exhibit very specific cysteine motifs in their amino acid sequence. Most chemokines have four characteristic cysteines, and depending on the motif displayed by the first two cysteines, they have been classified into CXC or alpha, CC or beta, C or gamma, and CX3C or delta chemokine classes [3]. The only exception to the four Cys rule is lymphotactin [4], which has only two Cys residues. Two disulfide bonds are formed between the first and third Cys and between the second and fourth. Recently, the molecular biology era brought with it the discovery of new chemokines that could activate and direct leukocytes in a coordinated fashion. At least 40 chemokines have been identified in humans. They mainly act on neutrophils, monocytes and lymphocytes, and play a pivotal role in host defense mechanisms [5]. The majority classes chemokines are CXC or CC chemokines [3]. Furthermore, the CXC or alpha subfamily has been divided into two groups depending on the presence of the ELR motif preceding the first cysteine: the ELR-CXC chemokines and the non-ELR-CXC chemokines. Those with the presence of the ERL motif, such as CXCL8 [interleukin (IL)-8] are potent neutrophil and T-lymphocyte chemoattractants [6, 7]. In contrast, CXCL10 [interferon-inducible protein 10 (IP-10)], which does not have an ERL motif, attracts T-lymphocytes [8] but not neutrophils. Human neutrophils perform a crucial role in the modulation of host immune response, by producing several kinds of cytokines, which regulate the functions of other immune cells such as T cells and B lymphocytes [9]. Importantly, Wiener's group [10] showed in their studies that mice vaccinated with gD DNA of HSV-1 and coinjected with human CXCL8 plasmid DNA exhibited a higher survival rate when compared to mice immunized with gD DNA alone. The possible explanation for this demonstrated protective effect was that CXCL8 directly influenced the CD4+ Th1 T-cell response.

The CC or beta subfamily, including macrophage inflammatory protein (MIP)-1alpha, MIP-1beta, RANTES (for regulated upon activation, normal T-cell expressed and secreted), and monocyte chemotactic protein-1 (MCP-1), promote chemotaxis and transendothelial migration of mononuclear cells [11, 12]. MIP-1 alpha, MIP-1 beta, and RANTES induce activation and proliferation of T cells [13] and of macrophages [14, 15], and MIP-1 alpha promotes Th1 cell differentiation [16, 17]. These CC-chemokines, play significant roles in granuloma formation [18, 19]. The ability of CC-chemokines to attract and activate T-cells and monocytes suggests that chemokines may have a role in modulating immune responses to bacterial infection. Importantly, it has been suggested that bronchial epithelial cell monolayer injury results in MCP-1 production, which initiates the repair process [20].

The chemokines are a family of secreted molecules known for their ability to regulate the trafficking of leukocytes from the circulation into tissues and lymphatic organs in the body [21-23] through interactions with a subset of seven-transmembrane domain, G_1 protein-coupled receptors [24, 25]. Based on the chemokine class they bind, the receptors have been named CXCR1, 2, 3, 4, and 5 (bind CXC chemokines); CCR1 through CCR9 (bind CC chemokines); XCR1 (binds the C chemokine, Lptn); and CX3CR1 (binds the CX3C chemokine)[26]. From the large group of chemokine receptors, CCR9 can be considered to be relatively mucosa specific. Chemokines are normally produced only transiently upon the action of an inducing stimulus and subsequently are downregulated rapidly. In addition, chemokines induce changes in target cell membrane protein composition, thus enhancing chemotaxis of leukocytes through the vascular endothelium into the inflamed tissues. This chapter does not aim to be a comprehensive review on chemokines; rather, we highlight those topics of chemokine biology with therapeutic potential in mycobacterial infection.

Chemokines in Innate Immunity

Recognition by the innate arm of the immune system leads to the release of chemokines, resulting in a state of inflammation. Under most circumstances, inflammation is a normal physiological and very beneficial response that protects us from invading microorganisms and infectious agents and precedes tissue repair. In recent years, chemokines have grown in importance, because they are involved in inflammation. For most infections, the innate immune system is sufficient to clear the infection. Innate immunity has been considered as the early and transitory first line of defense. Innate immune constituents, such as natural killer cells, dendritic cells, and macrophages, can promptly respond to bacteria [27].

Innate and adaptive immunity are two distinct but overlapping immune responses that play an integral role in orchestrating both short- and long-term protection against bacterial infection. Innate immune response occurs immediately following exposure to various pathogens and are initiated by the recognition of specific highly conserved structures of invading pathogens, called pathogen-associated molecular patterns (PAMPs)) [28, 29]. These PAMPs interact with pattern-recognition receptors known as PRRs [30, 31], which are expressed on a variety of cell types including macrophages, dendritic cells, endothelial cells and epithelial cells. The recognition of PAMPs by PRRs triggers an intracellular signaling pathway which culminates in induction of chemokines that leads to activation of adaptive immunity. Identification of a family of proteins similar to Toll in humans and mouse, namely the Toll-like receptors (TLRs), has revealed that they act as mammalian PRRs. To date, the TLRs represent the best-characterized pattern-recognition receptors. Specific TLR ligands interact with their cognate receptors and activate cells. Such activation results in maturation of cells with increased expression and secretion of chemokines. Recent intensive studies on signaling receptors of the innate immune system alert us to its importance for host defense. Innate immunity is equipped with various signaling receptors, including TLRs, that precisely recognize corresponding molecular patterns associated with microorganims and principally do not recognize host-derived intact components [32, 33]. In fact, TLR-mediated signaling pathways are essential for microorganism expulsion [34, 35]. Importantly, the innate immune recognition through TLRs provides an important approach to optimizing vaccines that prevent bacterial infection by employing activation of chemokines.

TLR in Innate Immunity

A new family of phylogenetically conserved receptors TLRs, has been discovered recently. TLRs are key components of the innate immune system that trigger antimicrobial host defense responses. These TLRs control several aspects of immunity, including chemokine gene expression [33]. Mammalian TLR proteins comprise a family of proteins that share sequence similarities with the *Drosophila* Toll receptor molecules [36]. TLRs have attracted particular interest because of their manifold functions in the regulation and linking of immune and inflammatory processes. TLRs have been identified as signaling receptors of the innate immune system recognizing a wide variety of molecular patterns for microorganisms [37]. These receptors have evolved as essential elements o fan alarm system suited for alerting the host to the presence of microorganims and for initiating anti-infective inflammatory response [38]. TLRs have emerged over the last 7-8 years as the major receptors involved in the recognition of self versus non-self pathogen-associated molecular patterns. At present, more than 12 mammalian TLRs and many of their ligands have been identified [39]. The first mammalian TLR member to be discovered was TLR4, and was shown to be the major receptor for endotoxins [lipopolysaccharide (LPS)], a major constituent of the cell wall produced by most Gram-negative bacteria [40]. In fact, LPS is recognized by the receptor complex of TLR4, mD-2, and CD14 [41-43], and initiating responses in both CD14-positive and CD14-negative cells in mice [44]. This response was also shown to be the case in humans. TLR4 was subsequently shown to be required for immune responses against mycobacteria [45]. Since then, at least 11 TLRs sharing homology with TLR4 have been identified in mammals, and each of these seems to recognize distinct pathogen-associated molecular patterns associated with different classes of microorganisms [46]. TLR2 alone is required for the response to a wide range of lipopeptides, and yeast carbohydrates [47]. It is also required for the recognition of peptidoglycan and other products derived from Gram-positive bacteria including *Staphylococcus aureus* [48]. In addition, it has been demonstrated that TLR2 recognizes lipoteichoic acid and bacterial lipoproteins abundant in the cell wall of Gram- positive and Gram-negative bacteria [49-51]. Viral double-stranded RNA stimulates TLR3 [52]. TLR5 is necessary for recognition of bacterial flagelins [53, 54]. TLR7 (in mice and humans) and TLR8 (in humans only) recognize nucleoside analogs such as resiquimoid or imiquimoid [55]. Human TLR8 is a signaling receptor for single-stranded RNA in viruses such as influenza virus and human immunodeficiency virus [56-58]. TLR9 recognizes specific hypomethylated (CpG) DNA sequences, and bacterial oligo DNA including unmethylated CpG motifs activates TLR9 [59, 60]. Humans (but not mice) express TLR10. Recently, it has been demonstrated that uropathogenic bacteria activate TLR11 [39]. Mice (but not humans) express TLR11,-12, and -13. It is important to note that currently, there is some confusion of nomenclature in that TLR11 has been called TLR12 in some reports, and vice versa [61].

In contrast to the above TLR members, TLR2 recognizes various types of PAMPs by forming the heterodimer with other TLR members. These TLRs have been shown to form functional pairs that can change their specificity or influence their signaling ability [62]. In fact, it has been demonstrated that pairing TLR2 with TLR1 results in the recognition of soluble factors released by *Neisseria meningitis* [63]. In addition, TLR2 together with TLR1 identifies bacterial lipoprotein exemplified by triacyl lipopeptide, such as 19 kDa lipoprotein purified from *Mycobacterium tuberculosis* [64], while TLR2 in association with TLR6 recognizes diacyl lipoprotein, such as macrophage-activating lipoprotein of 2 kDa (MALP-2) derived from *Mycoplasma fermentans* [65, 66]. Moreover, TLR2 needs to pair with TLR6 to recognize peptidoglycan [67].

Another important observation has been that in addition to exogenous ligands, TLRs are also involved in the recognition of endogenous or 'self' ligands. Therefore, the intrinsic host-derived molecules can also activate TLR4. Heat-shock proteins (HSP) 60, and 70 conserved proteins acting as intracellular molecular chaperones, were demonstrated to activate TLR4 to provoke immune responses [68, 69]. However, some reports have suggested that the role of TLR4 in the recognition of HSP60 and HSP70 may be due to contamination of these preparations with bacterial products, indicating the caution with which such observations should be interpreted [70]. Importantly, the recognition of HSP by TLR2 generates beta-chemokines which function as important adjuvants enhancing adaptive immunity [71].

In consistent with their role of pathogen recognition, TLRs are mainly expressed in cell types involved in the first line of defense, such as macrophages, dendritic cells, mucosal epithelial cells and endothelial cells [72]. Distinct subsets of these cells seem also to express different TLRs that enable them to induce specific types of immune responses to specific pathogens.

The structure of a representative TLR (TLR3) has recently been solved through X-ray crystallography, reveling that TLR3 is a dimeric protein composed of two horseshoe-shaped subunits that stack together side by side [73]. The cytoplasmic domains of TLR1 and TLR2 have also been crystallized and their structures solved [74, 75]. It is widely believed that all the TLRs are dimeric, with some homodimeric and others heterodimeric. All TLRs are single-spanning transmembrane proteins with ectodomains largely composed of leucine-rich repeats (LRRs), and with a cytoplasmic domain largely composed of a carboxy-terminal Toll-interleukin-1 (IL-1) receptor (TIR) domain. TLRs can be grouped into families according to the types of ligands they recognize. All TLRs possess amino-terminal leucine-rich repeats, which are responsible for the recognition of PAMPs, and also possess a carboxy-terminal TIR domain, which is required for initiating intracellular signaling [76].

At least some TLRs depend on other proteins to signal effectively [77]. In the case of the TLR2/TLR6 heterodimer, CD36 is known to participate in signaling events [66, 67]. Importantly, it has been demonstrated an essential role of CD14 in the TLR2/TLR6 heterodimer response, since mutations that eliminate CD14 partly impair sensing of all TLR2/TLR6 ligands [78].

TLR in Chemokine Gene Expression

Recognition of pathogen-associated molecular patterns by TLRs results in the activation of signaling events that induce the expression of effector molecules such as chemokines, and costimulatory molecules, which control the activation of antigen-specific and non-specific adaptive immune responses. However, different TLR ligands exert distinct but sometimes also overlapping biological effects, which can be attributed to the activation of unique and common signaling pathways leading to the induction of unique and common sets of genes. Thus, we discuss the potential involvement of TLRs and their signaling pathway in regulating chemokines network in bacterial diseases and their potential value as novel therapeutic targets.

Signal transduction pathways activated by TLRs have continued to be a major focus of research for investigators interested in the initiation of innate immune responses and the induction of chemokines during infection. Signal transduction involves the transmission of biochemical information from one part of the cell to another. Biochemical details of the major signaling pathways have now been obtained, and the specific signaling pathways activated by different TLRs are being elucidated. The recognition by TLRs of PAMPs triggers the TIR domain-dependent association with adapters containing the TIR domain [79]. In general, two major pathways are activated by TLRs [80]. The first of these culminates in the activation of the nuclear factor-kB (NF-kB), which acts as a master switch for inflammation, regulating the transcription of many genes that encode proteins involved in immunity and inflammation. The second leads to activation of several protein kinases, such as mitogen-activated protein kinases (MAPKs) [81], phosphatidylinositol-3′-kinase, and Jun amino-terminal kinase (JNK), which also participate in increased transcription and regulate the stability of mRNAs that contain AU repeats. With the exception of TLR3, it appears that all TLRs activate NF-kB and MAP kinases via a pathway that involves IL-1 receptor associated kinase (IRAK). A key issue concerns specificity because, although all TLRs can activate NF-kB and MAP kinases, there are differences in the ultimate gene expression profile that results from the activation of individual TLRs. Evidence exists to indicate a role for AP-1, often involving cooperative interactions with NF-kB, in the regulation of chemokine genes other than CCL5 such as CXCL8 gene activation in RSV-stimulated A549 cells and CCL2 gene activation in IL-1beta-stimulated endothelial cells [82, 83].

The fact that all TLRs signal via shared adapter molecules does not seem to limit the variety of immunological responses TLR signaling can induce [84]. It is important to note that although the intracellular domains of all TLR members have a common domain, namely the TIR domain [85, 86], their extracellular domains are quite different. Upon stimulation with ligands for TLRs, the intracellular TIR domain provides a platform for recruitment a signal adapter molecule, namely the myeloid differentiation factor 88 (MyD88) [87]. Subsequently, IRAK is recruited onto the platform on the basis of the interaction of IRAK and MyD88 [88]. Then, phosphorylated IRAK is dissociated from the receptor and associated with another signaling adapter molecule, tumor necrosis factor receptor-associated factor 6 (TRAF6) [89], eventually leading to phosphorylation of transforming growth factor-ß-activated kinase 1 (TAK1) [90]. TAK1 then phosphorylates complex of inhibitor of NF-kB (IkB) kinase (IKK) which induces ubiquitination-induced degradation of IkB, promoting the translocation of NF-kB to the nucleus [91]. TAK1 also activates MAPK pathways. Subsequently, corresponding gene expression is accomplished. Indeed, TLR signal transduction leads to the expression of several proteins with important roles in the innate immune response to microorganisms; these proteins include proinflammatory cytokines, costimulatory proteins, inducible nitric oxide synthase, and chemokines [77]. Indeed, TLRs can signal via four adapter proteins MyD88, Mal (MyD88 adapter-like), TRIF (a TIR domain-containing adapter inducing IFN-beta/TIR-containing adapter molecule-1), and TRAM (TRIF-related adapter molecule) [92-94]. MyD88 is important for signaling by all TLRs characterized thus far with the possible exception of TLR3 [95]. MyD88 was the first adaptor to be described and, similar to TLRs, it has a TIR domain. Without the adapter protein MyD88, which is required for all signal transduction by many of the TLRs, mice frequently succumb to bacterial infections. In addition, it has been demonstrated hat when both MyD88 and the adapter TRIF are absent, mice are even more susceptible to infection [96]. In fact, TLRs use different combinations of TIR domain-containing adapters to activate distinct transcription factors such as NF-kB, IRF3, IRF5 and IRF7, and therefore exhibit several biological responses. It has been shown that upon ligand binding, TLR family members recruit the adapter molecule MyD88 through homotypic interactions with a TIR domain that MyD88 contains in its C-terminus [97, 98]. The pathway bifurcates and results in the activation of the TRAF-6 and MAPKs. TRAF-6 subsequently activates the transcription factor NF-kB [89], and activation of NF-kB seems to be important for the expression of multiple genes that include chemokines such as MIP-1, and presumably secreted RANTES [99]. In fact, several studies have demonstrated that activation of NF-kB seems also to be important for the expression of multiple genes that include chemokines such as CXCL8 [100-103]. It was found that although MyD88 was essential for MAPK and NF-kB activation in response to TLR2 ligand [104], in response to TLR4 ligand LPS, MAPK, and NF.kB, activation appeared only transiently delayed but otherwise normal. These signal transduction pathways through individual TLRs make an essential contribution to the recognition of infectious organisms.

Moreover, TRIF has been shown to be unique among these adapters in its ability to activate IRF3 and IRF7 [105, 106]. This leads to the induction of a distinct set of genes that include CXCL10, and RANTES [107]. In fact, double-stranded activates NF-kappaB and IRFs and these transcription factors activate transcription of the RANTES promoter and its protein expression in airway epithelial cells [108]. Interestingly, IRF1 activation interact with TNF-alpha to regulate RANTES gene expression in human airway smooth muscle cells [109]. Furthermore, several experimental studies with TLR knockout mice have demonstrated that TLRs are involved in the production and release of immunomodulatory cytokines [110, 111]. With respect to this, the chemokine pattern alter TLR signaling differs depending on the bacterial stimulus used and the cell type examined [112, 113], and chemokines control recruitment of leukocyte populations through specific receptors that are differentially expressed by certain cellular types [114]. The following section detail some studies with diverse cellular types that address the signaling pathways through which TLRs generally signal to activate transcription of chemokine genes from mycobacteria infected cells. These pathways can differ in varying cell mycobacterial systems. At present, little is known about the role and interaction of TLRs with the adaptive immune system during infection of cells with mycobacteria. Although the link between TLR signaling and tuberculosis is far from being elucidated, several new studies report protection from mycobacterial diseases induced by bacteria signaling through TLR2, and TLR4 [115,116]. Understanding the specificity of the human chemokine response and exploring the intracellular signaling pathways that relate to *M.bovis* BCG are critical for defining the mechanisms responsible for host defense during tuberculosis. The demonstration of the major role that chemokines play in mycobacterial diseases and the successful targeting of chemokines in the clinic has fueled research onto the mechanisms that may be inducing chemokine gene expression.

Role of Chemokines in Mycobacterial Infection

The hallmark of *Mycobacterium*-induced pathology is characterized by an inflammatory response culminating in the formation of a granuloma [117]. In most healthy persons, acquired immunity, mediated by T cells, controls but not eradicate *M.tuberculosis* infection [118, 119]. Protective immunity against human tuberculosis requires productive infection of the host by *M.bovis* BCG, and this infection induces release of various chemoattractants. The early migration of activated monocytes/macrophages to the site of mycobacterial inflammation is an important step in the control of such infection [120]. Chemokine functions in leukocyte migration and organization of inflammatory reactions are recognized [121, 122].

Increased chemokine secretion has been observed in monocytes and polymorphonuclear granulocytes after infection with *M.tuberculosis* [123, 124]. In addition, it has been demonstrated that *M.tuberculosis*–induced chemokine secretion from macrophages is regulated by different factors, such as host cell surface molecules [116], and mycobacterial cell wall components [125]. Mycobacterial infection frequently induces significant expression of chemokines, including the CC chemokine subfamily members. These CC-chemokines, particularly RANTES, are potent leukocyte activators and chemoattractants, which have an important role in granuloma formation [126, 127]. In fact, it has been reported that murine macrophages infected in vitro with *M.tuberculosis* display induction of mRNAs for MIP-1 alpha, MIP-2, and MCP-1 [128], and that *M.tuberculosis* infection of human cells induces production of proteins for MIP-1 alpha, RANTES, and MCP-1 [129, 130]. Furthermore, it has been demonstrated that addition of anti-MCP-1 increased IL-12p40 levels in cultures of *M.tuberculosis*–stimulated monocytes, suggesting that this CC-chemokine is associated with increased susceptibility to pulmonary tuberculosis [131]. In addition, our results provided evidence that *M.bovis* BCG stimulates human monocytes to produce CC-chemokines, and that mobilization of intracellular Ca^{2+}, and the CD40 molecule are critical for the induction of RANTES [132]. The ability of CC-chemokines to attract not only monocytes/macrophages, but also T-cells suggests that this CC-chemokine class may has an important role in modulating immune responses to *M.tuberculosis* infection [133]. At present, it has been reported that CCL3, CCL4, and CCL5 are released by human alveolar macrophages upon infection with *M.tuberculosis*, and that all cell types responded to *M.tuberculosis* with CCL3 secretion. In addition, it has been shown that *M.tuberculosis* increased CCL2 secretion in monocytes, whereas multinucleate giant cell secreted CCL2 constitutively [134]. Tuberculous granulomas contain cell types potentially recruited by CCL5. Furthermore, it has been demonstrated that macrophages from human tuberculous lymph node granulomas express both CCL5 protein and gene [135], and that anti-CCL5 antibodies decrease pulmonary granuloma lesion size in *M.bovis* BCG strain-infected mice, suggesting a functional role for CCL5 in murine mycobacterial granulomas [136]. It has also been demonstrated that *M.tuberculosis* HSP70 stimulates human macrophages to produce CC-chemokines [137]. On the other hand, the current literature indicates that mycobacterial lipoarabinomannan induction of CC-chemokines from human cells is reduced by anti-tumor necrosis factor (TNF)-alpha antibody [138]. Our results, however, showed that the regulation of *M.bovis* BCG-induced CC-chemokines was not exclusively dependent of TNF-alpha, since the addition of neutralizing antibody to TNF-alpha failed to block secretion of MIP-1alpha and MIP-1beta, indicating that there are clearly additional mechanisms beyond TNF-alpha in determining *M.bovis*-induced CC-chemokine secretion [132].

During infection, mycobacteria induces increased expression of not only the CC chemokine subfamily members, but also of the CXC chemokine subfamily members, such as CXCL10 and CXCL8 [139, 140]. The following section detail studies with diverse mycobacterial systems that address the role of CXC chemokines in the regulation of mycobacterial infection. Recruitment of phagocytes to infected sites requires a chemokine gradient. CXCL10, in addition to its chemotactic properties, is also involved in the stimulation of natural killer and T-cell migration in *M.tuberculosis* infection [141]. CXCL10 secretion has been induced in *M.tuberculosis*-stimulated macrophages and constitutively in multinucleate giant cells [134]. We have demonstrated that following interaction with *M.bovis,* the human monocytes produced significant amounts of CXCL10, and that the type I

IFN may play an important role to regulate the expression of CXCL10 in *M.bovis* BCG infection [142].

CXCL8 is known to be involved in inflammatory conditions following infection by *M.tuberculosis* [143]. CXCL8 is not only involved in the recruitment of inflammatory cells to the sites of infection but also facilitates the elimination of microorganisms by increasing the efficiency of bactericidal activity by enhancing nonoxidative mechanisms [144]. It is clear that the attraction of inflammatory cells to tissue is controlled by chemokines such as CXCL8. However, in addition to chemotaxis, CXCL8 activates neutrophils, inducing degranulation, calcium mobilization, shape change, respiratory burst, upregulation of receptors and changes in phagocytic activity [145]. In neutrophils, CXCL8 enhances killing of *M.tuberculosis* [144], and neutrophils themselves secrete CXCL8 upon stimulation with *M.tuberculosis* [130]. *In vivo*, neutrophils are recruited to sites of infection after administration of *M.tuberculosis* or *M.bovis* BCG [146]. CXCL8 is an important chemokine which also stimulates chemotactic action in T cells, and basophils in tuberculosis [147]. Elevated CXCL8 concentrations are present in bronchoalveolar lavage fluid [148], pleural fluid [149], and plasma of tuberculosis patients and in *M.tuberculosis*-infected human tissue. Recently, it has been demonstrated that *M.tuberculosis* stimulation of multinucleate giant cells increased CXCL8 secretion [134]. At a cellular level, phagocytosis of *M.tuberculosis* by monocytic cells is an important stimulus to CXCL8 secretion [129]. In addition, we reported that human monocytes infected with *M.bovis* induce the secretion of CXCL8 [100,101]. In fact, CXCL8 is produced not only by inflammatory cells such as monocytes/macrophages, but also by human epithelial cells after exposure to *M.tuberculosis* and *M.bovis* [140]. Given the fact that the majority of mycobacteria infect via mucosal surfaces, immunity at that site is clearly of major importance. TNF-alpha and IFN-gamma have been shown to enhance RANTES gene expression in airway epithelial cells [150]. Human pleural fibroblasts play an important role in pleural responses when activated by TNF-alpha to release MCP-1 and CXCL8 [151]. In fact, during pleural inflammation, pleural mesothelial cells regulate the influx of neutrophils into the pleural space by polar production of CXCL8 [152]. Cocktail approaches involving chemokines, such as CXCL8 and RANTES, when given mucosaly, also appear to be effective in enhancing recruitment and boosting mucosal responses to a level sufficient for protection from mycobacteria challenge. Invasion of *M.tuberculosis* into the epithelium was shown to stimulate the secretion of CXCL8 [103]. Since it has been reported that the respiratory epithelium is a primary target for mycobacteria infection [153], we have also investigated the induction of CXCL8 by human epithelial cells infected with *M.bovis* BCG [140, 154]. Release of chemokines by human epithelial cells, which are considered the major cellular source of chemokines in the lung, can be selectively regulated by Th2-derived cytokines [155]. Our data demonstrated that *M.bovis* BCG-infected human epithelial dells can have an active role in a local inflammatory immune response via the secretion of CXC chemokines which can be selectively regulated by Th2-derived cytokines [156]. *M.bovis* BCG has a consistent record of protection against severe forms of primary tuberculosis in children. However, *M.bovis* BCG has poor efficacy against adult tuberculosis [157,158]. Several chemokines are now being studied for their potential use as adjuvants in antimycobacterial immune responses, providing a rational basis for optimization of vaccine potency by engineering the expression of the relevant ligands in recombinant strains of BCG [159]. Little is known, however, about the cellular signaling pathways by which mycobacterial antigens mediate CXC-chemokine induction. In this section, we reviewed the dependency on TLRs for mycobacterial-induced CXC-chemokine production. Several studies have chronicled in detail the interaction of *M.tuberculosis* secreted antigens and TLRs. The key information from these studies is that the 19 kDa lipoprotein and lipid derivatives of *M.tuberculosis* mainly interact with TLR2 to induce a predominantly proinflammatory cellular response [64]. Recent research has revealed that soluble and cell wall-associated mycobacterial factors from *M.bovis* BCG are capable of mediating activation via distinct TLR molecules [160]. Two mycobacterial cell wall glycolipids, lipoarabinomannan and mannosylated phosphatidylinositol (PIM) from *M.bovis* BCG, activate cells trhough TLR2 and TLR4 [161]. Furthermore, these authors reported that *M.bovis* BCG induces TNF-alpha secretion from human cells via TLRs [161]. Moreover, it is well known that TLR1 and TLR6 that heterodimerize with TLR2 have also been implicated in the recognition of mycobacterial antigens [64]. Molecular studies on the binding of mycobacterial components to the TLRs have implicated the involvement of these receptors in mycobacteria-induced proinflammatory signaling in human cells [162, 163], suggesting that TLRs are essential for the induction of protective immune response to human tuberculosis. Although it is clear that Th1 responses can confer resistance to acute tuberculosis, precisely what type of host immunity is capable of effectively controlling latent, chronic tuberculosis is poorly understood. Early studies by Jones and colleagues [164] demonstrated that Ara-LAM, and mannosylated phosphatidylinositol, isolated from *M.tuberculosis*, stimulated ERK 1/2 phosphorylation and activation of the transcription factors NF-kB and AP-1 in a murine cell line. This stimulation of the macrophage was apparently mediated through ligation of the TLR2. In mice, binding sites for the transcription factor NF-kB in the TLR2 promoter have been implicated in the induction of TLR2 mRNA expression by *M.avium* [165]. Mycobacterial induction of TLR2 signaling by lipoarabinomanan and phosphatidylinositol mannosides has been identified [166]. In vivo, it has been shown the transient defect in chemokine expression and uncontrolled bacteria in TLR2-deficient mice after low-dose infection with *M.tuberculosis* kurono strain [167]. In short-term, acute model of *M.bovis* BCG infection TLR2-deficient mice showed a 10-fold increase of pulmonary bacterial load 2 weeks postinfection, with a reduced adaptive response [168]. In addition, reduced TNF production in TLR2-deficient macrophages stimulated with *M.bovis* BCG has been reported, indicating that TLR2 is essential in macrophage proinflammatory cytokine such as TNF in response to *M.bovis* BCG particulate or soluble fractions. In addition, specific blocking of TLR4 resulted in a reduction in TNF-alpha production, indicating a substantial TLR4 involvement in *M.tuberculosis*-induced TNF-alpha secretion [169]. It has been shown that LPS induce neutrophil signaling leading to CXCL8 production through TLR2 and TLR4 [170]. Recently, it has been demonstrated the role of ERK, p38 and NF-kB signal transduction in regulation of LPS-induced TLR2, TLR4 and TLR9 gene expression [171]. Importantly, the inhibition of TLR2 significantly, and that of TLR4 partially, decreased the ERK1/2 activation induced by the *M. bovis* BCG when neutralizing antibodies to TLR2 or TLR4 were applied to human epithelial cells, and the levels of CXCL8 secretion and mRNA were much more markedly reduced by function blocking of both TLR2 and TLR4, indicating that the *M. bovis* BCG, acting through both TLR2 and TLR4, induces the activation of the ERK1/2 MAPK pathway, which in turn play a major role in CXCL8 secretion (our unpublished data). Future success in this area will most likely stem from creative strategies utilizing the most effective combination of inflammatory cytokines and chemokines that can hopefully elicit sufficiently protection against mycobacterial infection.

CONCLUSION

The past few years have witnessed an explosive growth in the discovery of new chemokines. In the near future it should be possible to gain further insights into the hyerarchical organization of the chemokine system, which is an essential requirement for any therapeutic exploitation. Chemokines produced during mycobacterial infections augment the host's innate resistance while shaping the ultimate adaptive immune response. Both RANTES and CXCL8 have important chemotactive effects through the activation of cells, and at least for CXCL8, enhanced killing of *M.tuberculosis*. In human tuberculosis, targeting of chemokines induced by the interaction of mycobacteria with TLRs may provide the basis for novel therapies, since the available evidence suggests that widespread vaccination strategy agains tuberculosis will be to combine different vaccine candidates in suitable adjuvant. In fact, the concept of using certain chemokines as natural adjuvants for human vaccines has been applied recently, and the use of these in combination and with other activators of cellular function might lead to novel and powerful approaches to mycobacterial infections. We now have a more complete understanding of the signal transduction pathways by receptors involve in induction of chemokine production, which allows the generation of models of how chemokines may be involves in mycobacteiral infection. TLRs are expressed in chronic inflammatory sites and can regulate several aspects of immunity and inflammation that include immune responses against mycobacteria, but most importantly chemokine gene expression. Therefore, although these studies are still in their early days, an important role of TLRs and TLR signaling in the mycobacterial diseases is plausible, and it has raised the interest for the mycobacterial therapies of the future.

ACKNOWLEDGEMENTS

This work was supported by grant 20060190 from the Coordinación General de Posgrado e Investigación de el IPN (CGPI). PMS is a COFAA, EDI, and SNI fellow.

REFERENCES

[1] Yoshie, O., Imai, T., & Nomiyama, H. 2001. Chemokines in immunity. *Adv Immunol* **78**: 57-110.

[2] Baggiolini, M., Dewald, B., & Moser, B. 1997. Human chemokines: an update. *Annu Rev Immunol* **15**: 675-705.

[3] Zlotnik, A., & Yoshie, O. 2000. Chemokines: a new classification system and their role in immunity. *Immunity.* **12**:121-127.

[4] Butcher, E.C., & Picker, L.J. 1996. Lymphocyte homing and homeostasis. *Science* **272**:60-66.

[5] Zlotnik, A. 2006. Involvement of chemokine receptors in organ-specific metastasis. *Contrib. Mirobiol.* **13**: 191-199.

[6] Taub, D.D., Anver, M., Oppenheim, J.J., Longo, D.L. & Murphy, W.J. 1996. T lymphocyte recruitment by interleukin-8 (IL-8). IL-8-induced degranulation of

neutrophils releases potent chemoattractants for human T lymphocytes both in vitro and in vivo. *J. Clin. Invest.* **97**: 1931-1941.

[7] Baggiolini, M., Walz, A. & Kunkel, S.L. 1989. Neutrophil-activating peptide-1/interleukin 8, a novel cytokine that activates neutrophils. *J. Clin. Invest.* **84**:1045-1049.

[8] Taub, .D., Lloyd, A.R., Conlon, K., Wang, J.M., Orlando, J.R., Harada, A., Matsushima, K., Kelvin, D.J., & Oppenheim, J.J. 1993. Recombinant human interferon-inducible protein 10 is a chemoattractant for human monocytes and T lymphocytes and promotes T cell adhesion to endothelial cells. *J Exp Med* **177**: 1809-1814.

[9] Mukaida, N., Harada, A., & Matsushima, K. 1998. Interleukin-8 (IL-8) and monocyte chemotactic and activating factor (MCAF/MCP-1), chemokines essentially involved in inflammatory and immune reactions. *Cytokine Growth Factor Rev* **9**: 9-23.

[10] Kim, J., Yang, S., Dentchev, T., Dang, K., & Weiner, D. 2000. Chemokine gene adjuvants can modulate immune responses induced by DNA vaccines. *J Interferon Cytokine Res* **20**: 487-498.

[11] Roth, S., Carr, M., & Springer, T. 1995. C-C chemokines, but not the C-X-C chemokines, interleukin-8 and interferon-gamma inducible protein-10, stimulate transendothelial chemotaxis of T lymphocytes. *Eur J Immunol* **25**: 3482-3488.

[12] Taub, D., Conlon, K., Lloyd, A., Oppenheim, J., & Kelvin, D. 1993. Preferential migration of activated CD4+ and CD8+ T cells in response to MIP-1 alpha and MIP-1 beta. *Science* **260**: 355-358.

[13] Taub, D., Turcovski-Corrales, S., Key, M., Longo, D., & Murphy, W. 1996. Chemokines and T lymphocytes activation: beta-chemokines costimulate human T lymphocyte activation in vitro. *J Immunol* **156**: 2095-2103.

[14] Lima, M., Zhang, Y., & Villalta, F. 1997. Beta-chemokines that inhibit HIV-1 infection of human macrophages stimulate uptake and promote destruction of *Trypanosoma cruzi* by human macrophages. *Cell Mol Biol* **43**: 1067-1076.

[15] Fahey, T., Tracey, K., Tekamp-Olson, P., Cousens, L., Jones, W., Shires, G., Cerami, A., & Sherry, B. 1992. Macrophage inflammatory protein-1 modulates macrophages function. *J Immunol* **148**: 2764-2769.

[16] Karpus, J., & Kennedy, K. 1997. MIP-1alpha and MCP-1 differentially regulate acute and relapsing autoimmune encephalomyeliltis as well as Th1/Th2 lymphocyte differentiation. *J Leukoc Biol* **62**: 681-687.

[17] Karpus, J., Luckas, N., Kennedy, K., Smith, W., Hurst, S., & Barret, T. 1997. Differential CC chemokine-induced enhancement of T helper cell cytokine production. *J Immunol* **158**: 4129-4136.

[18] Chensue, S., Warmington, K., Allenspach, E., Lu, B., Gerard, C., Kunkel, S., & Lakacs, N. 1999. Differential expression and cross-regulatory function of RANTES during mycobacterial (type 1) and schistosomal (type 2) antigen elicited granulomatous inflammation. *J Immunol* **163**: 165-173.

[19] Kurashima, K., Mukaida, M., Fujimura, M., Yasui, M., Nakazumi, Y., Matsuda, T., & Matsushima, K. 1997. Elevated chemokine levels in bronchoalveolar lavage fluid of tuberculosis patients. *Am J Respir Crit Care Med* **155**: 1474-1477.

[20] Lundien, M.C., Mohammed, K.A., Nasreen, N., Tepper, R.S., Hardwick, J.A., Sanders, K.L., Van Horn, R.D., & Antony, V.B. 2002. Induction of MCP-1 expression in airway

epithelial cells: role of CCR2 receptor in airway epithelial injury. *J Clin Immunol* **22**: 144-152.

[21] Baggiolini, M. 1998. Chemokines and leukocyte traffic. *Nature* **392**: 565-568.

[22] Moser, B., & Loetscher, P. 2001. Lymphocyte traffic control by chemokines. *Nat. Immunol.* **2**:123-128.

[23] Baggiolini, M., & Loetscher, P. 2000. Chemokines in inflammation and immunity. *Immunol. Today.* **21**: 418-420.

[24] Thelen, M. 2001. Dancing to the tune of chemokines. *Nat. Immunol.* **2**: 129-134.

[25] Murdoch, C., & Finn, A. 2000. Chemokine receptors and their role in inflammation and infectious diseases. *Blood* **95**: 3032-3043.

[26] Rossi, D. & Zlotnik, A. 2000. The biology of chemokines and their receptors. *Annu. Rev. Immunol.* **18**: 217-242.

[27] Hume, D., & Hertzog, P. 2006. Innate Imminty. *Curr Opin Immunol* **18**: 1-2.

[28] Medzhitov, R., & Janeway, C.A. 1997. Innate immunity: the virtues of a nonclonal system of recognition. *Cell* **91**:295-298.

[29] Medzhitov, R., & Janeway, C.A. 1997. Innate immunity: impact on the adaptive immune response. *Curr Opin Immunol* **9**:4-9.

[30] Hoffmann, J.A., Kafatos, F.C., Janeway, C.A., & Ezekowitz, R.A. 1999. Phylogenetics perspectives in innate immunity. *Science* **284**:1313-1318.

[31] Lemaitre, B., Nicolas, E., Michaut, L., Reichhart, J.M. & Hoffman, J.A. 1996. The dorsoventral regulatory gene cassette spatzle/Toll/cactus controls the potent antifungal response in *Drosophila* adults. *Cell* **86**:973-983.

[32] Janeway, C.A., & Medzhitov, R. 2002. Innate immune recognition. *Annu Rev Immunol* **20** :197-216.

[33] Takeda, K., Kaisho, T., & Akira, S. 2003. Toll-like receptors. *Annu Rev immunol* **21** : 335-376.

[34] Seki, E., Tsutsui, H., Tsuji, N.M., Hayashi, N., Adachi, K., Nakano, H., Futatsugi-Yumikura, S., Takeuchi, O., Hoshino, K., Akira, S., Fujimoto, J., & Nakanishi, K. 2002. Critical roles of myeloid differentiation factor 88-dependent proinflammatory cytokine release in early phase clearance of *Listeria monocytogenes* in mice. *J. Immunol.* **169**:3863-3868.

[35] Edelson, B.T., & Unanue, E.R. 2002. MyD88-dependent but Toll-like receptor 2-independent innate immunity to *Listeria:* no role for either in macrophage listericidal activity. *J. Immunol.* **169**:3869-3875.

[36] Rock, F.L., Hardiman, G., Timans, J.C., Kastelein, R.A., & Bazan,J.F. 1998. A family of human receptors structurally related to Drosophila Toll. *Proc. Natl. Acad. Sci. USA* **95**: 588-593.

[37] Takeda, K., & Akira, S. 2003. Toll receptors and pathogen resistance. *Cell Microbiol* **5**: 143-153.

[38] Iwasaki, A., & Medzhitov, R. 2004. Toll-like receptor control of the adaptive immune responses. *Nat Immunol* **5**: 987-995.

[39] Zhang, D., Zhang, G., Hayden, M.S., Greenblatt, M.B., Bussey, C., Flavell, R.A., & Ghosh, S. 2004. A Toll-like receptor that prevents infection by urophatogenic bacteria. *Science* **303**:1522-1526.

[40] Raetz, C.R.H., & Whitfield, C. 2002. Lipopolysaccharide endotoxins. *Annu Rev Biochem.* **71**: 634-700.

[41] Poltorak, A., He, X., Smimova, L., Liu, M.Y., Van Huffel, C., Du, X., Birdwell, D., Alejos, E., Silva, M., Galanos, C., Freudenberg, M., Ricciardi-Castagnoli, P., Layton, B., & Beutler, B. 1998. Defective LPS signaling in C3H/Hej and C57BL/10ScCr mice: mutations in Tlr4 gene. *Science* **282**:2085-2088.

[42] Qureshi, M.H., Zhang, T., Koguchi, Y., Nakashima, K., Okamura, H., Kurimoto, M., & Kawakami, K. 1999. Combined effects of IL-12 and IL-18 on the clinical course and local cytokine production in murine pulmonary infection with *Cryptococcus neoformans. Eur. J. immuno.* **29**:643-649.

[43] Nagai, Y., Akashi, S., Nagafuku, M., Ogata, M., Iwakura, Y., Akira, S., Kitamura, T., Kosugi, A., Kimoto, M., & Miyake, K. 2002. Essential role of MD-2 in LPS responsiveness and TLR4 distribution. *Nat Immunol.* **3**: 667-672.

[44] Perera, P-Y., Mayadas, T.N., Takeuchi, O., Akira, S., Zaks-Zilberman, M., Goyert, S.M. & Vogel, S.N. 2001. CD11b/CD18 acts in concert with CD14 and Toll-like receptor (TLR) 4 to elicit full lipopolysacharide and taxol-inducible gene expression. *J Immunol* **166**:574-581.

[45] Means, T., Wang, S., Lien, E., Yoshimura, A., Golenbock, D., & Fenton, M. 1999. Human Toll-like receptors mediate cellular activation by *Mycobacterium tuberculosis. J Immunol* **163**: 3920-3927.

[46] Medvedev, A.E., Lentschat, A., Wahl, L.M., Golenbock, D.T., & Vogel, S.N. 2002. Dysregulation of LPS-induced Toll-like receptor 4-MyD88 complex formation and IL-1 receptor associated kinase 1 activation in endotoxin-tolerant cells. *J Immunol* **169**: 5209- 5216.

[47] Redecke, V., Hacker, H., Datta, S.K., Fermin, A., Pitha, P.M., Broide, D.H., & Raz, E. 2004. Cutting edge: activation of Toll-like receptor 2 induces a Th2 immune response and promotes experimental asthma. *J Immunol* **172**:2739-2743.

[48] Lien, E., Sellati, T.J., Yoshimura, A., Flo, T.H., Rawadi, G., Finberg, R.W., Carroll, J.D., Espevik, T., Ingalls, R.R., Radolf, J.D., & Golenbock, D.T. 1999. Toll-like receptor 2 functions as a pattern recognition receptor for diverse bacterial products. *J Biol Chem* **274**: 33419-33425.

[49] Schwandner, R., Dziarski, R., Wesche, H., Rothe, M., & Kirschning, C.J. 1999. Peptidoglycan- and lipoteichoic acid-induced cell activation is mediated by Toll-like receptor 2. *J. Biol Chem* **274**: 17406-17409.

[50] Yoshimura, A., Lien, E., Ingalls, R.R., Tuomanen, E., Dziarski, R., & Golenbock, D. 1999. Cutting edge: recognition of Gram-positive bacterial cell wall components by the innate immune system occurs via Toll-like receptor 2. *J Immunol* **163**: 1-5.

[51] Re, F., & Strominger, J.L. 2001. Toll-like receptor 2 (TLR2) and TLR4 differentially activate human dendritic cells. *J Biol Chem* **276**: 37692-37699.

[52] Alexopoulou, L., Holt, A.C., Medzhitov, R., & Flavell, R.A. 2001. Recognition of double stranded RNA and activation of NF-kB by Toll-like receptor 3. *Nature* **413**: 732- 738.

[53] Gewirtz, A.T., Navas, T.A., Lyons, S., Godowski, P.J., & Madara, J.L. 2001. Bacterial flagellin activates basolaterally expressed TLR5 to induce epithelial proinflammatory gene expression. *J. Immunol.* **167**: 1882-1885.

[54] Hayashi, F., Smith, K.D., Ozinsky, A., Hawn, T.R., Yi, E.C., Goodlett, D.R., Eng, J.K., Akira, S., Underhill, D.M., & Aderem, A. 2001. The innate immune response to bacterial flagellin is mediated by Toll-like receptor 5. *Nature* **410**: 1099-1103.

[55] Jurk, M., Heil, F., Vollmer, J., Schettler, C., Krieg, A.M., Wagner, H., Lippford, G., & Bauer, S. 2002. Human TLR7 or TLR8 independently confer responsiveness to the antiviral compound R-848. *Nat Immunol* **3**:499.

[56] Hemmi, H., Kaisho, T., Takeuchi, O., Saro, S., Sanjo, H., Hoshino, K., Horiuchi, T., Tomizawa, H., Takeda, K., & Akira, S. 2002. Small antiviral compounds activate immune cells via TLR7 MyD88-dependent signaling pathway. *Nat Immunol.* **3**: 196- 200.

[57] Heil, F., Hemmi, H., Hochrein, H., Ampenberger, F., Kirschning, C., Akira, S., Lipford, G., Wagner, H., & Bauer, S. 2004. Species-specific recognition of single- stranded RNA via Toll-like receptor 7 and 8. *Science.* **303**: 1526-1529.

[58] Diebold, S.S., Kaisho, T., Hemmi, H., Akira, S., & Reis e Sousa, C. 2004. Innate antiviral responses by means of TLR7-mediated recognition of single-stranded RNA. *Science* **303**:1529-1531.

[59] Krieg, A.M. CpG motifs in bacterial DNA and their immune effects. *Annu Rev Immunol.* **20**:709-760.

[60] Hemmi, H., Takeuchi, O., Kawai, T., Kaisho, T., Sato, S., Sanjo, H., Matsumoto, M., Hoshino, K., Wagner, H., Takeda, K., & Akira, S. 2000. A Toll-like receptor recognizes bacterial DNA. *Nature* **408**:740-745.

[61] Yarovinsky, F., Zhang, D., Anderson, J.F., Bannenberg, G.L., Serhan, C.N., Hayden, M.S., Hieny, S., Sutterwala, F.S., Flavell, R.A., Ghosh, S., & Sher, A. 2005. TLR11 activation of dendritic cells by a protozoan profiling-like protein. *Science* **308**: 1626- 1629.

[62] Medzhitov, R., Preston-Hurlburt, P., & Janeway, C.A.Jr., 1997. A human homologue of the *Drosophila* Toll protein signals activation of adaptive immunity. *Nature* **388**:394-397.

[63] Woods, J.P., Frelinger, J.A., Warrack, G., & Cannon, J.G. 1998. Mouse genetics locus *Lps* influences susceptibility to *Neisseria meningitidis* infection. *Infect Immun* **56**: 1950- 1955.

[64] 64. Noss, E.H., Pai, R.K. Sellati, T.J., Radolf, J.D., Belisle, J., Golenbock, D.T., Boom, W.H., & Harding, C.V. 2001. Toll-like receptor 2-dependent inhibition of macrophage class II MHC expression and antigen processing by 19-kDa lipoprotein of *Mycobacterium tuberculosis. J Immunol* **167**:910-918.

[65] Takeuchi, O., Sato, S., Horiuchi, T., Hoshino, K., Takeda, K., Dong, Z., Modlin, R.L. & Akira, S. 2002. Role of Toll-like receptor 1 in mediating immune response to microbial lipoproteins. *J Immunol* **169**:10-14.

[66] Takeuchi, O., Kaufmann, A., Grote, K., Kawai, T., Hoshino, K., Morr, M., Muhlradt, P.F. & Akira, S. 2000. Preferentially the R-stereoisomer of the mycoplasmal lipopeptide macrophage-activating lipopeptide-2 activates immune cells though a Toll-like receptor 2- and MyD88-dependent signaling pathway. *J Immunol* **164**:554-557.

[67] Takeuchi, O., Kawai, T., Muhlradt, P.F., Morr, M., Radolf, J.D., Zychlinsky, A., Takeda, K., & Akira, S. 2001. Discrimination of bacterial lipoproteins by Toll-like receptor 6. *Int Immunol* **13**:933-940.

[68] Vabulas, R.M., Ahmad-Nejad, P., da Costa, C., Miethke, T., Kirschning, C.J., Hacker, H., & Wagner, H. 2001. Endocytosed HSP60s use Toll-like receptor 2 (TLR2) and TLR4 to activate the Toll/Interleukin-1 receptor signaling pathway in innate immune cells. *J Biol Chem* **276**: 31332-31339.

[69] Ohashi, K., Burkart, V., Flohé, S., & Kolb, H. 2000. Heat shock protein 60 is a putative endogenous ligand of the Toll-like receptor-4 complex. *J Immunol* **164**: 558-561.

[70] Da Costa, C.U., Wantia, N., Kirschning, C.J., Busch, D.H., Rodriguez, N., Wagner, H., & Miethke, T. 2004. Heat shock protein 60 from Chlamydia pneumoniae elicits an unusual set of inflammatory responses via Toll-like receptor 2 and 4 in vivo. *Eur J Immunol* **134**: 2874-2884.

[71] Kirschning, C.J., & Schumann, R.R. 2002. TLR2: cellular sensor for microbial and endogenous molecular patterns. *Curr Top Microbiol Immunol* **270**: 121-144.

[72] Hayashi, F., Means, T.K., & Luster, A.D. 2003. Toll-like receptors stimulate human neutrophil function. *Blood* **102**:2660-2669.

[73] Choe, J., Kelker, M.S., & Wilson, I.A. 2005. Crystal structure of human Toll-like receptor 3 (TLR3) ectodomain. *Science* **309**: 581-585.

[74] Xu, Y., Tao, X., Shen, B., Horng, T., Medzhitov, R., Manley, J.L., & Tong, L. 2000. Structural basis for signal transduction by the Toll/interleukin-1 receptor domains. *Nature* 408: **111**-115.

[75] Dunne, A., Ejdeback, M., Ludidi, P.L., O'Neill, L.A., & Gay, N.J. 2003. Structural complementarity of Toll/interleukin-1 receptor identity regions in Toll-like receptors and the adaptors Mal and MyD88. *J Biol Chem* **278**: 41443-41451.

[76] O´Neill, L.A.J., Fitzgerald, K.A., & Bowie, A.G. 2003. The Toll/IL-1 receptor adapter family grows to 5 members. *Trends Immunol* **24**: 286-290.

[77] Akira, S., Takeda, K., & Kaisho, T. 2001. Toll-like receptors: critical proteins linking innate and acquired immunity. *Nat Immunol* **2**: 675-680.

[78] Matsuguchi, T., Masuda, A., Sugimoto, K., Nagai, Y., & Yoshikai, Y. 2003. JNK-interacting protein 3 associates with Toll-like receptor 4 and is involved in LPS-mediated JNK activation. *EMBO J* **322**: 4455- 4464.

[79] Yamamoto, M., Sato, S., Mori, K., Hoshino, K., Takeuchi, O., Takeda, K., & Akira, S. 2002. A novel Toll/IL-1 receptor domain-containing adapter that preferentially activates the IFN-beta promoter in the Toll-like receptor signaling. *J Immunol* **169**: 6668-6672.

[80] McGettrick, A.F., & O,Neill, L.A.J. 2004. The expanding family of MyD88-like adapters in TLR signal transduction. *Mol. Immunol.***41**:577-582.

[81] Means, T., Wang, S., Lien, E., Yoshimura, A., Golenbock, D., & Fenton, M. 2000. The biology of Toll-like receptors. *Cytokine Growth Factor Rev.* **11**:219-232.

[82] Moriuchi, H., Moriuchi, M., & Fauci, A.S. 1997. Nuclear factor-kappa B potently up-regulates the promoter activity of RANTES, a chemokine that blocks HIV infection. *J Immunol* **158**: 3483-3491.

[83] Foletta, V.C., Segal, D.H., & Cohen, D.R. 1998. Transcriptional regulation in the immune system: all roads lead to AP-1. *J Leukoc Biol* **63**:139-152.

[84] Bashir., M.E., Louie, S., Shi, H.N., & Nagler-Anderson, C. 2004. Toll-like receptor 4 signaling by intestinal microbes influences susceptibility to food allergy. *J Immunol* **172**: 6978-6987.

[85] Medzhitov, R., Preston-Hurburt, P., Kopp, E., Stadlen, A., Chen, C., Ghosh, S., & Janeway, C.A. 1998. MyD88 is an adaptor protein in the Toll/IL-1 receptor family signaling pathways. *Mol Cell* **2**: 253-258.

[86] O'Neill, I. 2000. The Toll/interleukin-1 receptor domain: a molecular switch for inflammation and host defense. *Biochem Sic Trans* **28**:557-563.

[87] Wesche, H., Henzel, W.J. Shillinglaw, W, Li, S., & Cao, Z. 1997. MyD88: an adapter that recruits IRAK to the IL-1 receptor complex. *Immunity* **7**:837-847.

[88] Cao, Z., Hanzel, W.J. & Gao, X. 1996. IRAK: a kinase associated with the interleukin-1 receptor. *Science* **271**: 1128-1131.

[89] Deng, L., Wang, C., Spencer, E., Yang, L., Braun, A., You, J., Slaughter, C., Pickart, C., & Chen, Z.J. 2000. Activation of the IkB kinase complex by TRAF6 requires a dimeric ubiquitin-conjugating enzyme complex and a unique polyubiquitin chain. *Cell* **103**:351-361.

[90] Wang, C., Deng, L., Hong, M., Akkaraju, G.R., Inoue, J., & Chen, Z.J. 2001. TAK1 is a ubiquitin-dependent kinase of MKK and IKK. *Nature* **412**:346-351.

[91] Musikacharoen, T., Matsuguchi, T., Kikuchi, T., & Yoshikai, Y. 2001. NF-kappaB and STAT5 play important roles in the regulation of mouse Toll-like receptor 2 gene expression. *J Immunol* **166**: 4516-4524.

[92] Fitzgerald, K.A., Palsson-McDermott, E.M., Bowie, A.G., Jefferies, C.A., Mansell, A.S., Brady, G., Brint, E., Dunne, A., Gray, P., Harte, M.T., McMurray, D., Smith, D.E., Sims, J.E., Bird, T.A., & O'Neil, L.A. 2001. Mal (MyD88-adaptor-like) is required for Toll-like receptor-4 signal transduction. *Nature* **413**: 78-83.

[93] Adachi, O., Kawai, T., Takeda, K., Matsumoto, M., Tsutsui, H., Sakagami, M., Nakanishi, K., & Akira, S. 1998. Targeted disruption of the MyD88 gene results in loss of IL-1 and IL-18-mediated function. *Immunity* **9**: 143-150.

[94] Yamamoto, M., Sato, S., Hemmi, H., Sanjo, H., Uematsu, S., Kaisho, T., Hoshino, K., Takeuchi, O., Kobayashi, M., Fujita, T., Takeda, K., & Akira, S. 2002. Essential role for TIRAP in activation of the signalling cascade shared by TLR2 and TLR4. *Narute* **420**: 324-329.

[95] Sato, S., Takeuchi, O., Fujita, T., Tomizawa, H., Takeda, K., & Akira, S. 2002. A variety of microbial components induce tolerance to lipopolysacharide by differentially affecting MyD88-dependent and –independent pathways. *Int Immunol* **14**:783-791.

[96] Kawai, T., Adachi, O., Ogawa, T., Takeda, K., & Akira, S. 1999. Unresponsiveness of MyD88-deficient mice to endotoxin. *Immunity.* **11**:115-122.

[97] Fitzgerald, K.A., McWhirter, S.M., Faia, K.L., Rowe, D.C., Latz, E., Golenbock, D.T., Coyle, A.J., Liao, S.M., & Maniatis, T. 2003. IKKepsilon and TBK1 are essential components of the IRF3 signaling pathway. *Nat Immunol* **4**: 491-496.

[98] Fitzgerald, K.A., Rowe, D.C., Barnes, B.J., Caffrey, D.R., Visintin, A., Latz, E., Monks, B., Pitha, P.M., & Golenbock, D.T. 2003. LPS-TLR4 signaling to IRF-3/7 and NF-kappaB involves the toll adapters TRAM and TRIF. *J Exp Med* **198**: 1043-1055.

[99] Nelson, P.J., Kim, H.T., Manning, W.C., Goralski, T.J., & Krensky, A.M. 1993. Genomic organization and transcriptional regulation of the RANTES chemokine gene. *J Immunol* **151**: 2601-2612.

[100] Méndez-Samperio, P., Palma, J., & Vazquez, A. 2002. Signals involved in mycobactteria-induced CXCL-8 production by human monocytes. *J Interferon Cytokine Res* **22**: 189-197.

[101] Méndez-Samperio, P., Palma, J., & Vazquez, A. 2001. Roles of intracellular calcium and NF-kappaB in the Bacillus Calmette-Guerin-induced secretion of interleukin-8 from human monocytes. *Cell Immunol* **211**: 113-122.

[102] Natarajan, K., Singh, S., Buerke, T., Grunberger, D., & Aggarwal, B.B. 1996. Caffeic acid phenethyl ester is a potent and specific inhibitor of activation of nuclear transcription factor NF-kB. *Proc Natl Acad Sci USA* **93**: 9090-9095.

[103] Wickremasinghe, M.I., Thomas, L.H., & Friedland, J.S. 1999. Pulmonary epithelial cells are a source of IL-8 in the response to *Mycobacterium tuberculosis*: essential role of IL-1 from infected monocytes in a NF-kB-dependent network. *J Immnol* **163**: 3936- 3947.

[104] Dillon, S., Agrawal, A., Van Dyke, T., Landreth, G., McCauley, L., Koh, A., Maliszewski, C., Akira, S., & Pulendran, B. 2004. A Toll-like receptor 2 ligand stimulates Th2 responses in vivo, via induction of extracellular signal-regulated kinase mitogen-activated protein kinase and c-Fos in dendritic cells. *J Immunol* **172**: 4733- 4743.

[105] Oshiumi, H., Sasai, M., Shida, K., Fujita, T., Matsumoto, M., & Seya, T. 2003. TIR-containing adapter molecule (TICAM)-2, a bridging adapter recruiting to toll-like receptor 4 TICAM-1 that induces interferon-beta. *J Biol Chem* **278**: 49751-49762.

[106] Yamamoto, M., Sato, S., Hemmi, H., Uematsu, S., Hoshino, K., Kaisho, T., Takeuchi, O., Takeda, K., & Akira, S. 2003. TRAM is specifically involved in the Toll-like receptor 4-mediated MyD88-independent signaling pathway. *Nat Immunol* **4**: 1144- 1150.

[107] Akira, S. 2006. TLR signaling. *Curr Tpo Microbiol Immunol* **311**:1-16.

[108] Ieki, K., Matsukura, S., Kokubu, F., Kimura, T., Kuga, H., Kawaguchi, M., Odaka, M., Suzuki, S., Watanabe, S., Takeuchi, H., Schleimer, R.P., & Adachi, M. 2004. Double-stranded RNA activates RANTES gene transcription through co-operation of nuclear factor-kappaB and interferon regulatory factors in human airway epithelial cells. *Clin Exp Allergy* **34**:745-752.

[109] Huang, C.D., Ammit, A.J., Tliba, O., Kuo, H.P., Penn, R.B., Panettieri, R.A. Jr. & Amrani, Y. 2005. G-protein-coupled receptor agonists differentially regulate basal or tumor necrosis factor-alpha-stimulated activation of interleukin-6 and RANTES in human aiway smooth muscle cells. *J Biomed Sci.* **12**:763-776.

[110] Sugawara, I., Yamada, H., Li, C., Mizuno, S., Takeuchi, O., & Akira, S. 2003. Mycobacterial infection in TLR2 and TLR6 knockout mice. *Microbiol Immunol* **47**: 327-336.

[111] Geisel, R.E., Sakamoto, K., Russell, D.G.,& Rhoades, E.R. 2005. In vivo activity of released cell wall lipids of Mycobacterium bovis bacillus Calmette-Guérin is due principally to trehalose mycolates. *J Immunol* **174**: 5007-5015.

[112] Li, X., Gong, Z.Y., Li, H., Wei, Q., Shi, M., & Yang, Y.R. 2004. Study on Toll-like receptors expression and cytokine production induced by bacillus Calmette-Guerin in human bladder cancer cell. *Zhonghua Wai Ke Za Zhi* **42**: 177-181.

[113] Miyazaki, J., Kawai, K., Oikawa, T., Johraku, A., Hattori, K., Shimazui, T., & Akaza, H. 2006. Uroepithelial cells can directly respond to Mycobacterium bovis bacillus Calmette-Guerin through Toll-like receptor signaling. BJU Int. **97**: 860-864.

[114] Lukacs, N.W., Hogaboam, C.M., & Kunkel, S.L. 2005. Chemokines and their receptors in chronic pulmonary disease. *Curr Drug Targets Inflamm Allergy* **4**:313-317.

[115] Reiling, N., Holscher, C., Fehrenbach, A., Kroger, S., Kirschning, C.J., Goyert, S., & Ehlers, S. 2002. Cutting edge: Toll-like receptor (TLR)2- and TLR4-mediated pathogen

recognition in resistance to airborne infection with *Mycobacterium tuberculosis. J Immunol* **169**: 3480-3484.

[116] Lopez, M., Sly, L.M., Luu, Y., Young, D., Cooper, H., & Reiner, N.E. 2003. The 19-kDa Mycobacterium tuberculosis protein induces macrophage apoptosis through Toll-like receptor-2. *J Immunol* **170**: 2409-2416.

[117] Flynn, J.L., & Chan, J. 2001. Immunology of tuberculosis. *Annual Review of Immunology.* **19**:93-129.

[118] Kaufmann, S.H. 1993. immunity to intracellular bacteria. *Annu Rev Immunol* **11**: 129- 163.

[119] Kaufmann, S.H. 2001. How can immunology contribute to the control of tuberculosis? *Nat Rev Immunol* **1**: 20-30.

[120] Flynn, J.L., & Chan, J. 2005. What´s good for the host is good for the bug. *Trends Microbiol* **13**:98-102.

[121] Zlotnik, A. 2006. Chemokines and cancer. *Int J Cancer* **119**: 2026- 2029.

[122] Otto, G., Burdick, M., Strieter, R., & Godaly, G. 2005. Chemokine response to febrile urinary tract infection. *Kidney Int* **68**: 62-70.

[123] Riedel, D.D., & Kaufmann, S.H. 1997. Chemokine secretion by human polymorphonuclear granulocytes after stimulation with *Mycobacterium tuberculosis* and lipoarabinomannan. *Infect Immun* **65**: 4620-4623.

[124] Lin, Y.G., Gong, J., Zhang, M., Xue, W., & Barnes, P.F. 1998. Production of monocyte chemoattractants protein 1 in tuberculosis patients. *Infect Immun* **66**: 2319- 2322.

[125] Mohamed, K.A., Nasreen, N., Ward, M.J., Muharak, K.K., Rodriguewz-Panadero, F., & Antony V.B. 1998. Mycobacterium-mediated chemokine expression in pleural mesothelial cells: role of C-C chemokines in tuberculosis pleurisy. *J Infect Dis* **178**: 1450-1456.

[126] Collins, H.L., & Kaufmann, S.H.E. 2001. The many faces of host responses to tuberculosis. *Immunology* **103**: 1-9.

[127] Bergeron, A., Bonay, M., Kambouchner, M., lecossier, D., Riquet, P., Soler, A., Hance, A., & Tazi, A. 1997. Cytokine patterns in tuberculosis and sarcoid granulomas : correlation with histopathologic features of the granulomatous response. *J Immunol* **159**: 3034-3043.

[128] Rhoades, E., Cooper, A., & Orme, I. 1995. Chemokine response in mice infected with *Mycobacterium tuberculosis. Infect Immun* **63**:3871-3877.

[129] Lin, Y., Zhang, M., & Barnes, P.F. 1998. Chemokine production by a human alveolar epithelial cell line in response to *Mycobacterium tuberculosis. Infect Immun* **66**: 1121- 1126.

[130] Kasahara, K., Sato, I., Ogura, K., Takeuchi, H., Kobayashi, K. & Adachi, M. 1998. Expression of chemokines and induction of rapid cell death in human blood neutrophils by *Mycobacterium tuberculosis. J Infect Dis* **178**:127-137.

[131] Flores-Villanueva, P.O., Ruiz-Morales, J.A., Song, C.H., Flores, L.M. Jo, E.K., Montano, M., Barnes, P.F., Selman, M., & Granados, J. 2005. A functional promoter polymorphism in monocyte chemoattractant protein-1 is associated with increased susceptibility to pulmonary tuberculosis. *J Exp Med* **202**: 1649-1658.

[132] Méndez-Samperio, P., Vazquez, A., & Ayala, H. 2003. Infection of human monocytes with *Mycobacterium bovis* BCG induces production of CC-chemokines. *Journal of Infection* **247**: 139- 147.

[133] Wickremasinghe, M.I., Thomas, L.H., O'Kane, C.M., Uddin, J. & Friedland, J.S. 2004. Transcriptional mechanisms regulating alveolar epithelial cell-specific CCL5 secretion in pulmonary tuberculosis. *J Biol Chem* **279**: 27199-27210.

[134] Zhu, X.W., & Friedland, J.S., 2006. Multinucleate giant cells and the control of chemokine secretion in response to Mycobacterium tuberculosis. *Clin Immunol* **120**: 10- 20.

[135] Sadek, M.I., Sada, E., Toossi, Z., Schwander, S.K., & Rich, E.A. 1998. Chemokines induced by infection of mononuclear phagocytes with mycobacteria and present in lung alveoli during active pulmonary tuberculosis. *Am J Respir Cell Mol Biol.* **19**: 513-521.

[136] Coelho, A.L., Hogaboam, C.M., & Kunkel, S.L. 2005. Chemokines provide the sustained inflammatory bridge between innate and acquired immunity. *Cytokine Growth Factor Rev* **116**: 553-560.

[137] Lehner, T., Bergmeier, L.A., Wang, Y., Tao, L., Sing, M., Spallek, R., & van der Zee, R. 2000. Heat shock protein generates beta-chemokines which function as innate adjuvants enhancing adaptive immunity. *Eur J Immunol* **30**: 594-603.

[138] Juffermans, N.P., Verboan, A., van Deventer, S.J., van Deurekom, H., Belisle, J.T., Ellis, M.E., Speelman, P., & van der Poll, T. 1999. Elevated chemokine concentrations in sera of human immunodeficiency virus (HIV) seropositive and HIV-seronegative patients with tuberculosis: a possible role for mycobacterial lipoarabinomannan. *Infect Immun* **67**: 4295-4297.

[139] Zhang, Y., Broser, M., Cohen, H., Bodkin, M., Law, K., Reibman, J., & Rom, W.N. 1995. Enhanced interleukin-8 release and gene expression in macrophages after exposure to *Mycobacterium tuberculosis* and its components. *The Journal of Clinical Investigation.* **95**: 586-592.

[140] Méndez-Samperio, P., Palma-Barrios, J., Vazquez-Hernandez, A., & Garcia- Martinez, E. 2004. Secretion of interleukin-8 by human-derived cell lines infected with *Mycobacterium bovis. Mediators of Inflammation.* **13**: 45-49.

[141] Orme, I.M., & Cooper, A.M. 1999. Cytokine/chemokine cascades in immunity to tuberculosis. *Immunol Today* **20**: 307-312.

[142] Méndez-Samperio, P., Trejo, A., & Miranda, E. 2004. Role of type I interferon in the bacillus Calmette-Guérin-induced expression of CXCL10 from human monocytes. *Mediators of Inflammation.* **13**: 343- 348.

[143] Amexia, C., & Friedland, J.S. 2001. Down-regulation of interleukin-8 secretion from *Mycobacterium tuberculosis*-infected monocytes by interleukin-4 and -10 but not by interleukin-13. *Infect immun* **69**:2470- 2476.

[144] Nibbering, P.H., Pos, O., Stevenhagen, A., & Van Furth, R. 1993. Interleukin-8 enchances nonoxidative intracellular killing of *Mycobacterium fortuitum* by human granulocytes. *Infect Immun* **61**:3111-3116.

[145] Baggiolini, M. 1995. Activation and recruitment of neutrophil leukocytes. *Clin Exp Immunol* **101**: 5- 6.

[146] Fulton, S.A., Martin, T.D., Redline, R.W., & Boom H. W. 2000. Pulmonary immune responses during primary *Mycobacterium bovis*-Calmette-Guèrin bacillus infection in C57Bl/6 mice. *Am J Respir Cell Mol Biol* **22**: 333- 343.

[147] Pace, E., Gjomarkaj, M., Melis, M., Profita, M., Spatafora, M., Vignola, A.M., Bonsignore, G., & Mody, C.H. 1999. Interleukin-8 induces lymphocyte chemotaxis into

the pleural space. Role of pleural macrophages. *Am J Respir Crit Care Med.* **159**: 1592-1599.

[148] Casarini, M., Ameglio, F., Alemanno, L., Zangrilli, P., Mattia, P, Paone, G., Bisetti, A., & Giosue, S. 1999. Cytokine levels correlate with a radiologic score in active pulmonary tuberculosis. *Am J Respir Crit Care Med* **159**: 143-148.

[149] Yamada, Y., Nakamura, A., Hosoda, M., Kato, T., Asano, T., Tonegawa, K., & Itoh M. 2001. Cytokines in pleural liquid for diagnosis of tuberculous pleurisy. *Respir Med* **95**: 577-581.

[150] Casola, A., Henderson, A., Liu, T., Garofalo, R.P., & Brasier, A.R. 2002. Regulation of RANTES promoter activation in alveolar epithelial cells after cytokine stimulation. *Am J Physiol Lung Cell Mol Physiol* **283**: L1280-L1290.

[151] Loghmani, F., Mohammed, K.A., Nasreen, N., Van Horn, R.D., Hardwick, J.A., Sanders, K.L., & Antony, V.B. 2002. Inflammatory cytokines mediate C-C (monocyte chemotactic protein 1) and C-X-C (interleukin 8) chemokine expression in human pleural fibroblasts. *Inflammation* **26**: 73-82.

[152] Nasreen, N., Mohammed, K.A., Hardwick, J., Van Horn, R.D., Sanders, K.L., Doerschuk, C.M., Hott, J.W., & Antony, V.B. 2001. Polar production of interleukin-8 by mesothelial cells promotes the transendothelial migration of neutrophils: role of intercellular adhesion molecule-1. *J Infect Dis* **183**: 1638-1645.

[153] Thurlbeck, W.M. *1984.* Overview of the pathology of pulmonary emphysema in the human. *Clinics in Laboratory Medicine* **4**: 539-559.

[154] Méndez-Samperio, P., Trejo, A., & Miranda, E. 2005. *Mycobacterium bovis* BCG induces CXC chemokine ligand 8 secretion via the MEK-dependent signal pathway in human epithelial cells. *Cell Immunol* **234**: 9- 15.

[155] Fujisawa, T., Kato, Y., Atsuta, J., Tereda, A., Iguchi, K., Kamiya, H., Yamada, H., Nakajima, T., Miyamasu, M., & Hirai, K. 2000. Chemokine production by the BEAS-2B human bronchial epithelial cells: differential regulation of eotaxin, IL-8, and RANTES by TH2- and TH1-derived cytokines. *The Journal of Allergy and Clinical Immunology.* **105**: 126-133.

[156] Méndez-Samperio, P., Miranda, E., & Vázquez, A. 2006. Expression and secretion of CXCL-8 and CXCL-10 from *Mycobacterium bovis* BCG-infected human epithelial cells: Role of IL-4. *Mediators of Inflammation.* IN PRESS.

[157] Kaufmann, S.H., Baumann, S., & Nasser Eddine, A. 2006. Exploting immunology and molecular genetics for rational vaccine design against tuberculosis. *Int J Tuberc Lung Dis* **10**: 1068-1079.

[158] Orme, I.M. 2005. Tuberculosis vaccines : current progress. *Drugs* **65**:2437-2444.

[159] Kaufmann, S.H. 2006. Envisioning future strategies for vaccination against tuberculosis. *Nat Rev Immunol* **6**:699-704.

[160] von Meyenn, F., Schaefer, M., Weighardt, H., Bauer, S., Kirschning, C.J., Wagner, H., & Sparwasser, T. 2006. Toll-like receptor 9 contributes to recognition of Mycobacterium bovis Bacillus Calmette-Guerin by Flt3-ligand generated dendritic cells. *Immunobiology* **211**: 557-565.

[161] Tsuji, S., Matsumoto, M., Takeuchi, O., Akira, S., Azuma, I., Hayashi, A., Toyoshima, K., & Seya, T. 2000. Maturation of human dendritic cells by cell wall skeleton of Mycobacterium bovis Calmette-Guerin: involvement of toll-like receptors. *Infect Immun* **68**:6883- 6890.

[162] Underhill, D.M., Ozinsky, A., Smith, K.D., & Aderem, A. 1999. Toll-like receptor-2 mediates mycobacteria-induced proinflammatory signaling in macrophages. *Proc Natl Acad Sci* **96**: 14459- 14463.

[163] Heldwein, K.A., Liang, M.D., Andressen, T.K., Thomas, K.E., Marthy, A.M., Cuesta, N., Vogel, S.N., & Fenton, M.S. 2003. TLR2 and TLR4 serve distinct roles in the host immune response against *Mycobacterium bovis* BCG. *J Leukoc Biol* **74**: 277-286.

[164] Jones, B.W., Heldwein, K.A., Means, T.K., Saukkonen, J.J., & Fenton, M.J. 2001. Differential roles of Toll-like receptors in the elicitation of proinflammatory responses by macrophages. *Ann Rheum Dis* **60**: 6-12.

[165] Wang, T., Lafuse, W.P., & Zwilling, B.S. 2001. NF-kappaB and Sp1 elements are necessary for maximal transcription of Toll-like receptor 2 induced by *Mycobacterium avium*. *J Immunol* **167**: 6924- 6932.

[166] An, H., Yu, Y., Zhang, M., Xu, H., Qi, R., Yan, X., Liu, S., Wang, W., Guo, Z., Guo, J., Qin, Z., & Cao, X. 2002 Involvement of ERK, p38 and NF-kB signal transduction in regulation of TLR2, TLR4 and TLR9 gene expression induced by lipopolysaccharide in mouse dendritic cells. *Immunology* **106**: 38- 45.

[167] Means, T.K., Jones, B.W., Schromm, A.B., Schurtleff, B.A., Smith, J.A., Keane, J., Golenbock, D.T., Vogel, S.N., & Fenton, M.J. 2001. Differential effects of a Toll-like receptor antagonist on *Mycobacterium tuberculosis*-induced macrophage responses. *J Immunol* **166**: 4074- 4082.

[168] Abel, B., Thieblemont, N., Quesniaux, V.J., Brown, N., Mpagi, J., Miyake, K., Bihl, F., & Ryffel, B. 2002. Toll-like receptor 4 expression is required to control chronic *Mycobacterium tuberculosis* infection in mice. *J Immunol* **169**: 3155- 3162.

[169] Fenton, M.J. 2001. Different Toll-like receptor agonists induce distinct macrophage responses. *J Leukoc Biol* **69**: 1036- 1044.

[170] Neufert, C., Pai, R.K., Noss, E.H., Berger, M., Boom, W.H., & Harding, C.V. 2001. *Mycobacterium tuberculosis* 19-kDa lipoprotein promotes neutrophil activation. *J Immunol* **167**: 1542-1549.

[171] Sabroe, I., Prince, L.R., Jones, E.C., Horsburgh, M.J., Foster, S.J., Vogel, S.N., Dower, S.K., & White, M.K. 2003. Selective roles for Toll-like receptor (TLR)2 and TLR4 in the regulation of neutrophil activation and life span. *J Immunol* **170**: 5268- 5275.

In: Chemokine Research Trends
Editor: Leah R. Grinwald, pp. 29-46
ISBN: 978-60021-588-9

Chapter 2

Herpes Simplex Virus Type 1 Suppresses Type I Interferon-Induced Chemokine Production in Trigeminal Ganglion Cell Cultures

Bobbie Austin and Daniel J. J. Carr
Department of Ophthalmology, Microbiology and Immunology.
The University of Oklahoma Health Sciences Center, 608 Stanton L Young Blvd.,
Oklahoma City, OK 73104 USA

Abstract

Type I interferon (IFN) subtypes initiate the transcription of IFN-inducible gene products by binding to a shared receptor that is broadly expressed on the surface of most cells. Although the receptor for each subtype is communal, the specific downstream events initiated by each subtype are distinct. The purpose of this study was to compare and contrast the induction or augmentation of chemokine production by different classes of type I IFNs in HSV-1 infected and uninfected trigeminal ganglion cell cultures. Overall, the following observations were made: 1) CCL5 was induced by IFN-α1, -α4, and -α9; 2) CXCL10 production was induced by IFN-α1, -α4, -α5, and -β; 3) CCL2 production was augmented by IFN-α1, -α4, -α9, and -β. In all instances, HSV-1 infection was associated with a reduction in IFN-induced chemokine production. Abridged chemokine production was associated with suppression of cytokine signaling-3 expression rather than apoptosis. Collectively, the results suggest HSV-1 suppresses chemokine production by TG cells identifying another means by which the virus circumvents the host's immune response.

Introduction

Ocular herpes simplex virus type 1 (HSV-1) infection in mice often results in a robust inflammatory response with a massive recruitment of leukocytes into both the cornea and trigeminal ganglion (TG) [1-3]. Post-infection immune cell infiltration is associated with the expression of chemokines including CCL2, CCL3, CCL5, CXCL1, and CXCL10 [4-6]. Mice deficient in specific chemokines (e.g., CCL3) or antibody neutralization of chemokines (e.g., anti-CXCL10 IgG) dramatically reduces the influx of cells into the infected sites [7, 8]. This reduction in cellular infiltration often leads to a transient rise in HSV-1 titers that may favor the establishment of a latent infection. Accordingly, blocking the expression of chemokines should favor viral replication and spread in the host.

Although HSV-1 elicits a potent immune response in the murine host, the virus is still quite capable of replication and spread. The success of HSV-1 stems from its ability to evade both adaptive and innate immune responses as well as its ability to establish a latent infection in sensory ganglia. Mechanisms for evasion of the adaptive immune response include: 1) down regulation of major histocompatability complex (MHC) class I expression [9, 10]; 2) interference with the MHC class II processing pathway [11]; and 3) prevention of antibody-dependent cell cytotoxicity [12]. Likewise, HSV-1 has developed strategies to counter innate immunity as it relates to type I interferons (IFN) through viral-encoded proteins including $_{\gamma 1}34.5$, US11, and ICPO that antagonize IFN-inducible anti-viral pathways [13-16] or the type I IFN receptor Jak1/Stat1 signaling cascade [17]. These countermeasures along with the capacity to establish a latent infection in neurons of the sensory ganglion contribute toward the high prevalence of HSV-1 in the adult population worldwide [18].

Most of the morbidity associated with HSV-1 infection is due to the reactivation of the virus from latent-infected neurons typically in the TG [19]. The success of viral reactivation reportedly increases as a direct result of latent-infected neurons, and increased numbers of latent-infected neurons are associated with higher input virus [20]. These results imply that by suppressing the host immune response to acute infection, viral replication would go unchecked resulting in a larger virus burden capable of establishing latency providing such infection does not result in the death of the host. Previously, we have found IFN-α4 and IFN-β but not IFN-α5, -α6, or -α9 suppress HSV-1 replication in primary TG cell cultures in a dose- and RNase L-dependent fashion [21]. The current study expands upon these initial observations by investigating the Jak1/Stat1 signaling pathway following induction with type I IFN as well as the local chemokine responses to type I IFN in the presence and absence of HSV-1. Our results suggest HSV-1 significantly diminishes type I IFN-induced chemokine production by primary TG cells.

Materials and Methods

Virus and Cell Line

Vero cells and L929 cells originally obtained from the American Type Culture Collection (ATCC, Manassas, VA.) were cultured in RPMI-1640 (Invitrogen, Carlsbad, CA) or DMEM (for L929 cells, ATCC) supplemented with 10% fetal bovine serum (FBS, Invitrogen, Carlsbad, CA) and antibiotic/antimycotic solution (Invitrogen) at 37° C, 5% CO_2, and 95% humidity. HSV-1 stock (McKrae strain) was prepared as previously described [4]. Vesicular stomatitis virus (VSV) was a gift from Dr. Robert Fleishmann (Dept. Microbiology, UTMB).

Plasmid DNA Constructs

The description of the type I IFN containing plasmid constructs has previously been reported [22].

Primary TG Cells

Primary TG cell cultures were established by aseptically dissecting trigeminal ganglia from male and female C57BL/6 mice (Jackson Labs, Bar Harbor, ME), ages 6-8 weeks. Following dissection, TGs were placed in type IV and XI collagenase (Sigma, St. Louis, MO), diluted 1mg/ml in 1x phosphate buffered saline (pH 7.4), and incubated at 37° C, 5% CO_2, and 95%. To facilitate cellular dissociation, TGs were triturated every 20 min for 1 hr using a 1000 ul pipettor. At the end of the incubation period, the cells were washed twice with RPMI-1640 containing 10% FBS, and subsequently resuspended in TG growth medium (RPMI-1640 containing 10 ng/ml nerve growth factor 2.5s) (Collaborative Biomedical Products, Bedford, MA), antimycotic/antibiotic solution, and 10% FBS]. Collagenized cells were plated (1-3 x 10^5 cells/well) on 24-well microtiter plate wells (ISC BioExpress, Kaysville, UT), coated with laminin (Trevigen, Gaithersburg, MD) and type I collagen (Sigma). Following a 10-12 day incubation period at 37° C, 5% CO_2, and 95% humidity, the experiments using these cultures were initiated.

Primary Spleen Cells

Spleen cells obtained from C57BL/6 mice (male and female, 6-8 weeks of age) were treated with 0.84% NH_4CL to lyse red blood cells. Spleen cells were subsequently plated in wells of 24-well microtiter plates at a concentration of 1 x 10^6 cells/ml in RPMI-1640 medium containing 10% FBS and antimycotic/antibiotic solution. The cells were then exposed to type I IFN and/or HSV-1.

Generation of Type I IFNs

L929 cells were transfected with plasmid vector alone or plasmid vector containing type I IFNs including IFN-α1, -α4, -α5, -α6, -α9, or -β as previously described [21]. Twenty-four hr post-transduction, the supernatant from transfected cells was removed and assayed for the quantity of biologically-active IFN [21]. The supernatants were then stored at 4°C until use (within 2-3 weeks).

IFN Treatment and Infection of TG and Spleen Cells

TG or spleen cells were treated with 32.0 – 100.0 units of type I IFN, a combination of type I IFN, or one of the following controls: supernatant from plasmid, vector-transfected cells (no measurable IFN content), or media alone (DMEM containing antibiotic/antimycotic solution and 10% FBS). Twenty-four hr later (indicated as 0 time point), 200 ul media was transferred to 500 ul microcentrifuge tubes followed by assessment of chemokine/cytokine content by ELISA. The cells were subsequently infected with HSV-1 [McKrae strain, multiplicity of infection (MOI) = 0.5]. Twenty-four hr post-infection (p.i.) supernatants were collected and assayed for HSV-1 titer (for spleen cells) and chemokine/cytokine content was again analyzed by ELISA (indicated as 24 hr time point).

Chemokine/Cytokine ELISAs

The content of CCL2, CCL3, CCL5, CXCL10, TNF-α, and IFN-γ in the supernatant of TG and spleen cells incubated with or without type I IFNs and HSV-1 was determined by ELISA in duplicate using antibody pairs specific for each cytokine/chemokine following the manufacturer's protocol (R&D Systems, Minneapolis, MN). A standard curve using known concentrations of recombinant cytokine/chemokine was performed for each assay in duplicate. The concentration for each sample was extrapolated from the standard curve generated (correlation coefficient ranging from 0.900 – 0.9932 for seven points).

Measurement of Total Tyrosine Kinase Activity

TG cells were incubated with 32 units/ml of type I IFN for 30 min or two hrs. Following the incubation period, the cells were rinsed in ice-cold PBS (pH 7.4) and lysed in lysis buffer (50 mM Hepes buffer, pH 7.4, containing 0.1% Triton X-100, 10% glycerol, 1 mM dithiothreitol, 1mM sodium vanadate, and protease inhibitor cocktail) (Calbiochem, San Diego, CA) for 20 min. Lysates were centrifuged at 10,000 x g for 15 min at 4° C, and the supernatant was collected and assayed for protein content (Pierce, Rockville, IL). Tyrosine kinase activity was determined using a commercially available protein tyrosine kinase assay kit according to the manufacturer's instructions (Sigma). Epidermal growth factor receptor (of known activity) was used to generate a standard curve (correlation coefficient ranging from 0.92 –0.9600 for 5 points) used to extrapolate the quantity of tyrosine kinase for each unknown sample.

Western Blot Analysis

Thirty minutes to 2 hrs post IFN exposure, total cell protein was harvested following disruption of cells in sodium dodecyl sulfate (SDS) lysis buffer supplemented with a protease inhibitor cocktail containing 500 μM AEBSF, Hydrochloride, 150 nM aprotinin, 1 μM E-64 protease inhibitor, 0.5 mM disodium EDTA, and 1 μM hemisulfate leupeptin (Calbiochem, San Diego, CA). Protein was quantified with a BCA protein assay (Pierce, Rockville, IL) using a FL600 fluorescence plate reader (Bio-Tek Instruments Inc., Winooski, VT). Total cell protein (50 μg) was electrophoresed on 7.5% Tris-HCl Criterion precast gels (Bio-Rad, Hercules, CA) and semi-dry transferred to polyvinylidene difluoride (PVDF) membranes (Bio-Rad). PVDF membranes were blocked with 5% dry milk in 1x T-TBS [0.1% Tween, 0.1 M Tris (pH 7.4), 0.4 M NaCl] overnight at 4° C. Western detection was performed by incubation for 2 hr at room temperature on a shaking platform with the following primary antibodies: rabbit polyclonal anti-mouse phopho-Stat1 (Y701), Biosource, Camarillo, CA: rabbit polyclonal anti-mouse anti-Stat1 p91/p84 antibody, Santa Cruz Biotech. Inc., Santa Cruz, CA; rabbit polyclonal anti-β-actin antibody, Santa Cruz; or a goat polyclonal anti-SOCS3 antibody (Santa Cruz). The primary antibody was detected by a secondary incubation for 1 hour at room temperature using anti-rabbit or anti-goat antibody coupled to horseradish peroxidase (Santa Cruz). Horseradish peroxidase was detected using ECL Western blotting chemiluminescence detection reagents (Pierce), and images were obtained by exposing PVDF membranes to x-ray film. Resulting images were digitalized using the Bio-Rad 1000 image documentation system and software (Quantity One 4.0.3; Bio-Rad). In the case of SOCS-3 detection, IFN-treated TG cells were infected with HSV-1 (MOI = 0.5) 12 hr prior to assaying for SOCS-3 expression. To determine whether equivalent volumes of protein were loaded, membranes were stripped by submersion in stripping buffer (100mM 2-mercaptoethanol, 2% sodium dodecyl sulphate, 62.5 mM Tris-HCl, pH 6.7) and incubated at 50° C for a minimum of 30 minutes. Membranes were subsequently washed 2x for 10 minutes with large volumes of 1x T-TBS, blocked with 5% dry milk in 1x T-TBS, and a 2nd western detection was performed using antibodies to either Stat-1 or β-actin, proteins expected to be expressed at equivalent concentrations in all samples

Detection of Apoptotic Cells

TG cells (1 x 10^5 cells) were incubated 24 hr with 32 units/ml of type I IFN and then infected with HSV-1 (MOI = 0.5) for 6 or 12 hrs. Following the incubation period, the cells were fixed in 1% paraformaldehyde in PBS (pH 7.4) for 10 minutes, washed twice in PBS, post-fixed in pre-cooled ethanol:acetic acid (2:1) for 5 min at –20° C for 5 min, and then washed twice more in PBS. Endogenous peroxidase was then quenched with 3.0% hydrogen peroxide and assayed for DNA strand breaks using an apoptosis detection system according to the manufacturer's protocol (ApopTax, Chemicon International, Temecula, CA).

Statistics

Analysis of data was performed using one-way analysis of variance (ANOVA) and Tukey's post-doc t-test to determine statistically significant differences ($p<.05$) between the control and IFN-treated groups).

RESULTS

Expression of Phopho-Stat1 in TG Cells

Previously, we found that IFN-β and to a lesser extent, IFN-α4 suppress HSV-1 replication in primary TG cells relative to other type I IFNs including IFN-α1, -α5, -α6, and -α9 [21]. Even though IFN-α and -β species act through the same type I IFN receptor, differences may reside in signaling cascades (e.g. tyrosine kinases) activated by type I IFNs [23]. Therefore, as a means to address differences in anti-viral efficacy as a result of exposure to different type I IFNs, we initially measured total protein tyrosine kinase (PTK) activity in primary TG cell cultures following exposure to type I IFNs. Within 30 min post exposure, TG cells incubated in the presence of IFN-β but not other type I IFNs exhibited a 2-fold elevation in total PTK activity compared to control (vector)-treated TG cells (Fig. 1). By 2 hr post stimulation, IFN-α4, -α5, and -β stimulated cells all showed significant increases in total PTK activity compared to the control (vector)-treated TG cells (Fig. 1). By 2 hr post stimulation, IFN-α1, IFN-α4, and IFN-β stimulated cells also showed significant increases in total PTK activity compared to the control-treated cells (Fig. 1).

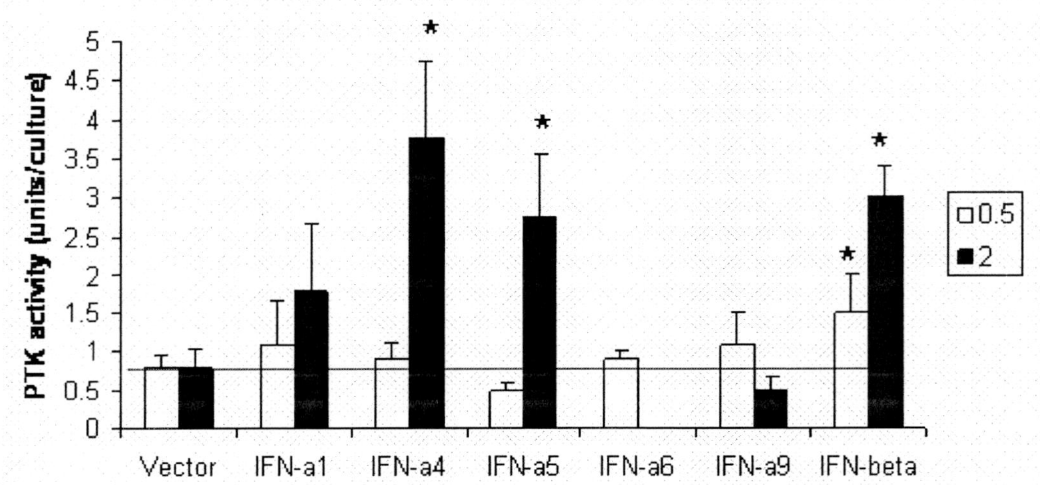

Figure 1. Total protein tyrosine kinase activity in IFN-treated TG cells in culture. Total protein tyrosine kinase (PTK) activity was measured in TG cells (1-3 x 10^5 cells) exposed to supernatants from vector or the indicated IFN transgene transfected L929 cells containing 32 units of biologically active IFN (except vector control). The supernatant from non-transfected cells served as the control for background PTK activity. Duplicate TG cell cultures/condition were assayed for PTK activity by ELISA at 30 min (0.5 hr) and 120 min (2 hr) intervals. The figure summarizes the results from two experiments. Bars represent the mean +/- SEM. Dotted line indicates the basal level of PTK activity in TG cells. *$p<.05$ comparing the IFN-treated to vector-treated cells for the 0.5 or 2 hr time point.

Since the principal signaling cascade involving PTK activity for anti-viral gene induction by type I IFNs is the Jak1/Stat1 pathway [23], analysis of Jak1 and Stat1 levels were assayed post-IFN exposure. There were no obvious differences in total Jak1 or phospho-Jak1 levels in the IFN-treated TG cells compared to the vector- or control-treated cells (data not shown). However, all type I IFN-treated TG cells possessed an increase in phospho-Stat1 expression with TG cell cultures exposed to IFN-α5, -α6, and -β expressing the greatest increase in phospho-STAT1 expression (Fig. 2). Total Stat1 levels were similar or modestly decreased with IFN-α6 compared to the control-treated cultures (Fig. 2). Collectively, these results suggest that IFN-α subtypes modify PTK activity at different strengths relative to IFN-β which shows a consistent increase in PTK and phospho-Stat1 levels. Since IFN-α4 and IFN-β suppresses HSV-1 replication in primary TG cells compared to other IFN-α species (which show no protective effect) [21], the current results also suggest alternative signaling pathways in addition to the Jak1/Stat1 cascade are contributory toward the anti-viral efficacy of IFN-α4.

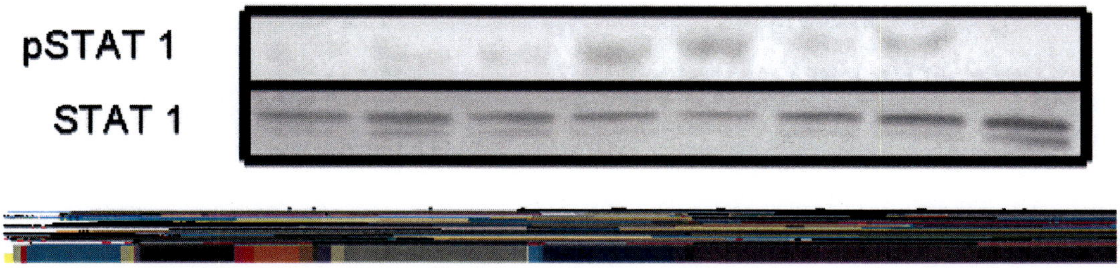

Figure 2. Phospho-Stat1 levels are elevated in IFN-treated TG cells. Total cell lysates (50 μg/lane) from IFN-treated (32 units) and non-treated (neg) or vector (Vec) supernatant-treated TG cells (1 x 10^5 cells, 30 min exposure) were electrophoresed on polyacrylamide gels, transferred to PVDF membranes, and used for Western blot analysis with polyclonal antibody specific for phospho-Stat1 or total Stat 1. This figure is representative of three experiments with similar outcomes.

Chemokine Expression by Primary TG Cells

Upon ocular infection with HSV-1, a significant rise in macrophages (F4/80$^+$), neutrophils (Gr-1$^+$), and CD11b$^+$ cells (expressed on macrophages, NK cells, neutrophils, and dendritic cells) in the TG has been reported [1]. Candidate chemokines that would facilitate leukocyte recruitment include CXCL10, CCL2, CCL3, and CCL5 among others. Therefore, we investigated the production of chemokines by TG cells in the presence and absence of type I IFN exposure prior to and after HSV-1 infection. Baseline (untreated, uninfected TG cell cultures) CCL2 levels from TG cell cultures approached one ng/ml whereas CXCL10 levels were barely detected (Fig. 3). In contrast, CCL5 was not detected in untreated, uninfected TG cell cultures (Fig. 3, baseline). Since type I IFN gene expression is found in the TG early during ocular HSV-1 infection [24], the impact of type I IFN on TG cell production of chemokines was investigated. TG cell cultures responded differently depending on the type I IFN species employed as a stimuli. Specifically, IFN-α1, -α4, -α9, and -β enhanced CCL2 production by TG cells whereas IFN-α1, -α5, and -α9 stimulated CCL5 production and IFN-α1, -α4, -α5, and -β enhanced CXCL10 production (Fig. 3). HSV-1 infection (24 hr post infection, baseline) suppressed baseline levels of CCL2 and stimulated CCL5 production ($p<.05$ comparing uninfected to infected levels) (Fig. 3). There was no significant change in baseline levels of CXCL10 in TG cell cultures following infection with HSV-1 (Fig. 3). In those cultures in which chemokine levels were enhanced by type I IFN, HSV-1 infection reduced the amounts to basal levels with the exception of IFN-α1 enhancement of CCL2 (Fig. 3). Although CCL3 levels were assessed, there was no detectable level prior to or after IFN exposure or HSV-1 infection (data not shown). In addition to chemokine production, we also investigated two prominent cytokines, IFN-γ and TNF-α, previously found to be expressed in the TG following ocular HSV-1 infection [25]. However, neither cytokine was detected in TG cell culture supernatants prior to or after exposure to type I IFN or HSV-1.

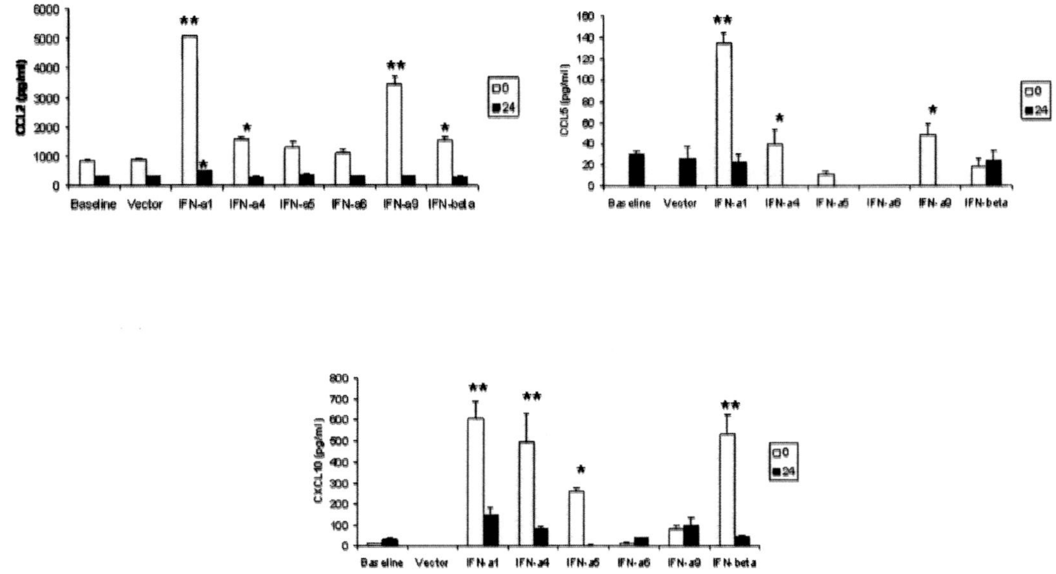

Figure 3. Chemokine secretion by IFN-stimulated TG cells. TG cells (1-3 x 10^5 cells/culture) were treated with supernatants from vector or the indicated IFN transgene transfected L929 cells containing 32 units of biologically active IFN. The supernatant from non-transfected cells served as the control for chemokine levels and did not differ from vector supernatant exposed cells. Following a 24 hr incubation period, 200 μl of the supernatant was removed for assaying chemokine/cytokine content (0) and the cells were infected with HSV-1 (MOI = 0.5). Twenty four hr post infection, the supernatant (24) was removed and assayed for chemokine content by ELISA. This figure is a summary of 4 experiments. Bars represent the mean +/- SEM. **p<.01, *p<.05 comparing the IFN-treated to vector- or basal levels for the respective chemokine.

Table 1. Additive Effect of Type I IFNs on Chemokine Production by Primary TG Cells[a]

Treatment	CCL2	CCL2 Post HSV-1	CCL5	CCL5 Post HSV-1
Vector	817 +/- 27[b]	381 +/- 17	0	113 +/- 33
IFN-α1	4000 +/- 119	589 +/- 31	484 +/- 9	107 +/- 34
IFN-β	3677 +/- 162	562 +/- 88	197 +/- 23	145 +/- 50
IFN-α + IFN-β	9527 +/- 794**	916 +/- 114*	766 +/- 54**	288 +/- 33*

[a] TG cells (1-3 x 10^5 cells/culture) were treated with supernatants from vector or the indicated IFN transgene transfected L929 cells containing 32 units of biologically active IFN. Following a 24 hr incubation period, 200 μl of the supernatant was removed for assaying chemokine/cytokine content, and the cells were infected with HSV-1 (MOI = 0.5). Twenty four hr post infection, the supernatant was removed and assayed for chemokine content by ELISA. [b]Bars represent the mean +/- SEM, n=3/group. **p<.01, *p<.05 comparing the IFN-α1 + IFN-β treated group to all other groups.

Since a previous report suggested the combination of type I IFNs act in concert to block viral replication [26], we investigated the anti-viral action of combining two efficacious type I IFNs, IFN-α1 and IFN-β against HSV-1 replication in primary TG cells. Although we found no additional effect of combining 32 units each of IFN-α1 and IFN-β above that compared to the anti-viral action of either type I IFN alone in measuring virus titers (data not shown), there was an additive effect in assessing chemokine levels. Specifically, combining 32 units of

IFN-α1 and IFN-β augmented the production of CCL2 and CCL5 by primary TG cells in comparison to either IFN-α1 or IFN-β treatment alone prior to of after infection with HSV-1 (Table 1). In contrast, there was no additive effect of the type I IFNs measuring CXCL10 production prior to or after HSV-1 infection (data not shown).

HSV-1 Elicits Up-Regulation of SOC-3

Since HSV-1 has been found to induce a pro-apoptotic state in T cells [27, 28] and was found to diminish IFN-induced or -enhanced chemokine levels in the current study, we investigated the incidence of apoptosis as a means to explain the reduction in chemokine secretion in IFN-treated TG cells. Although there was an increase in detectable levels of DNA fragmentation above control-treated cultures by 6 hr post infection, in some type I IFN-exposed cultures (IFN-α5, -α6, -α9, and -β), these levels remained extremely modest and did not differ from control-treated cultures by 12 hr post infection (Fig. 4). As an alternative explanation to viral-mediated chemokine suppression, suppressor of cytokine signaling (SOCS) proteins are induced by the Jak/Stat pathway and act to negatively regulate this

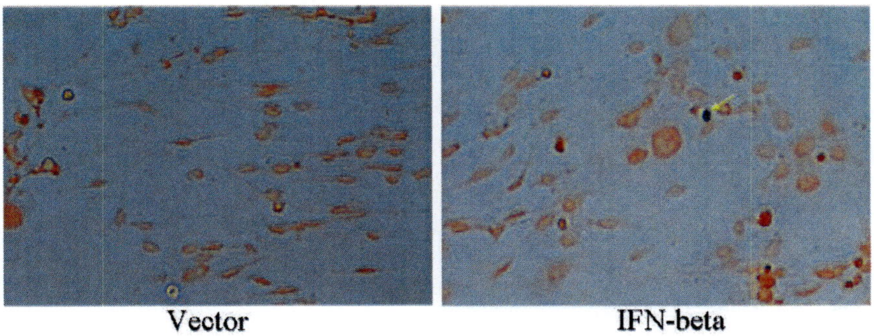

Time	Vect	IFN-a1	IFN-a4	IFN-a5	IFN-a6	IFN-a9	IFN-B	No Trt.
6 Hr	0.63 +/- 0.03	0.42 +/- 0.04	0.59 +/- 0.09	**1.31 +/- 0.02***	**1.93 +/- 0.04***	**1.70 +/- 0.19**	**1.81 +/- 0.01***	0.48 +/- 0.1
12 Hr	3.0 +/- 0.7	2.0 +/- 0.3	1.5 +/- 0.2	2.0 +/- 0.2	3.2 +/- 0.6	4.5 +/- 0.3	3.2 +/- 0.8	2.3 +/- 0.3

Figure 4. HSV-1 has a minimum impact on apoptosis in TG cells. TG cells (1-3 x 10^5 cells/culture) were treated with supernatants from vector or the indicated IFN transgene transfected L929 cells containing 32 units of biologically active IFN. Twenty four hr post treatment, the cells were infected with HSV-1 (MOI = 0.5). Six or twelve hr post infection, the cultures were fixed and assessed for apoptotic cells. Upper panel represents a typical profile at 6 hr post infection comparing vector supernatant- to IFN-β supernatant-treated TG cells. The yellow arrow highlights a cell undergoing apoptosis in the IFN-β treated group. The lower panel is a table summarizing the results of two experiments/time point each conducted in duplicate with 5-10 fields counted/well/treatment group with an average of 481 +/- 35 cells/field. *p<.05 comparing IFN-treated to vector supernatant-treated cells at 6 hr post infection.

pathway [29]. Moreover, HSV-1 has been found to suppress Jak1/Stat1 phosphorylation [17] and most recently, induce SOCS-3 expression in a human amnion cell line [30]. HSV-1 infection of TG cells resulted in a rise in SOCS-3 expression compared to non-infected cells (Fig. 5). The increase was not due to differences in total protein loaded as equivalent amounts of cell lysates were electrophoresed as indicated by similar β-actin levels (Fig. 5). Equally important, pre-treatment of TG cells with type I IFN did not modify HSV-1-mediated up-regulation of SOCS-3 expression (Fig. 5).

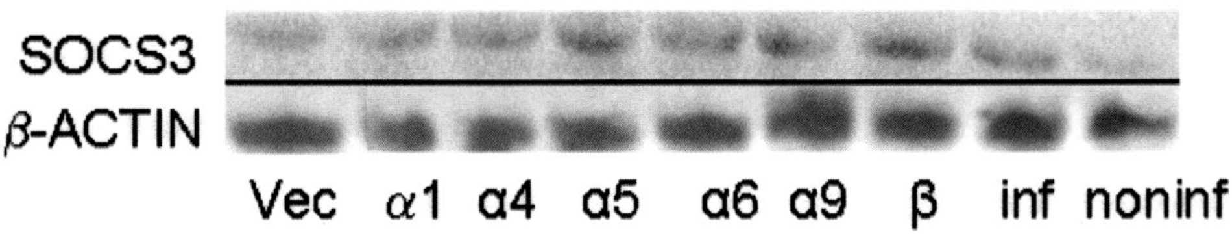

Figure 5. HSV-1 augments SOCS-3 expression by TG cells. Total cell lysates (50 μg/lane) from non-infected, infected but not pre-treated, or vector- or type I IFN supernatant pre-treated, HSV-1 infected TG cells were generated 12 hr post infection and electrophoresed on polyacrylamide gels. Following electrophoresis, the separated proteins were transferred onto PVDF membranes and used for Western blot analysis with polyclonal antibody specific for SOCS-3 or β-actin protein. This figure is a representative of three experiments with similar outcomes.

HSV-1 Augments Chemokine and Cytokine Expression by Spleen Cells

In primary TG cells infected with HSV-1, we observed very little induction or increase in chemokine expression (comparing baseline at 0 to 24 hr) and a suppression of type I IFN-induced chemokine expression. To determine if this effect was universally applicable, spleen cells were exposed to type I IFN and subsequently infected with HSV-1. Unlike primary TG cells [21], spleen cells exposed to 32 units of type I IFNs (IFN-α species or IFN-β) showed no resistance to HSV-1 infection compared to control-treated cells (Fig. 6). At 100 units of type I IFN, both IFN-α1 and IFN-β suppressed HSV-1 replication in HSV-1-infected spleen cells (Fig. 6). This dose of IFN was also evaluated for induction of chemokine/cytokine secretion in the presence or absence of HSV-1 infection. In the absence of stimulation, spleen cells produced little to no detectable chemokines or cytokines (Fig. 7). Specifically, CCL2 and CCL3 were below detectable levels (15 pg and 20 pg for CCL2 and CCL3 respectively) and CCL5, CXCL10, and TNF-α were less than 100 pg/ml (Fig. 7). Consistent with what was found in primary TG cell cultures, spleen cells exposed to type I IFNs had varied responses for measurements of selective chemokines or cytokine (Fig. 7). Specifically, all type I IFN species induced or augmented CCL2, CCL3, or CCL5 secretion by spleen cells. However, only IFN-β augmented CXCL10 secretion and IFN-α6 and IFN-α9 enhanced TNF-α production (Fig. 7). Following infection with HSV-1, chemokine and TNF-α production by non-stimulated (non IFN-treated) spleen cells was elevated. Unlike post-infected TG cells, type I IFN-stimulated cells produced equivalent levels or more of the chemokines with the exception of CCL2. In the case of CCL2, type I IFN-induced levels dropped to that obtained in the vector supernatant-treated group post HSV-1 infection (Fig. 7). Relative to TNF-α production, there was a significant increase in this pro-inflammatory cytokine by non-stimulated spleen cells (Fig. 7). In the presence of most type I IFNs, the levels of TNF-α produced by HSV-1-infected spleen cells remained similar to those of the vector, supernatant-treated group with the exception of cells treated with IFN-α1 or IFN-β. Both IFN-α1 and IFN-β reduced the level of TNF-α production by the HSV-1 infected spleen cells. There were no detectable levels of IFN-γ in non-stimulated or type I IFN-stimulated spleen cells prior to or after HSV-1 infection (data not shown). Collectively, these results demonstrate the tissue-specific nature of type I IFN action as it relates to HSV-1 infection and cytokine and chemokine production in spleen cells versus primary TG cells.

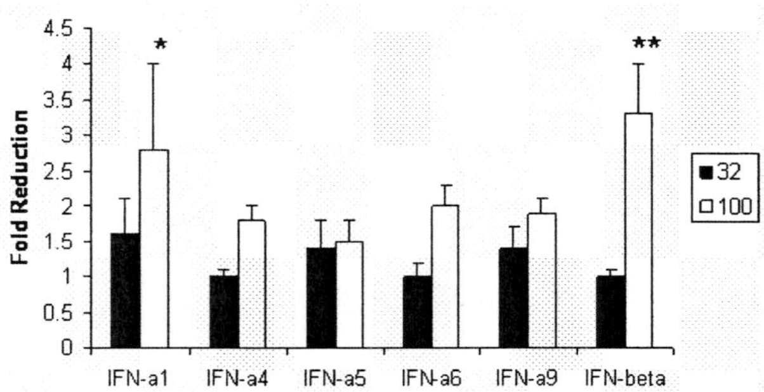

Figure 6. IFN-α1 and IFN-β suppress HSV-1 replication in spleen cells. Spleen cells (1 x 10^6 cells/well) were treated with supernatants from vector or the indicated IFN transgene transfected L929 cells containing 32 or 100 units of biologically active IFN. Twenty-four hr post treatment, the spleen cells were infected with HSV-1 (MOI = 0.5). Following a 24 hr incubation period, the supernatant was collected and viral titers were determined by plaque assay. This figure is a summary of 3-5 experiments for each treatment. **p<.01, *p<.05 comparing the IFN-treated to vector supernatant-treated control. Bars represent mean fold inhibition +/- SEM compared to the vector-treated group (mean viral titer of 1,562,500 +/- 152,000 plaque forming units/ml). Similar levels of HSV-1 were recovered in non-treated versus vector supernatant-treated cells.

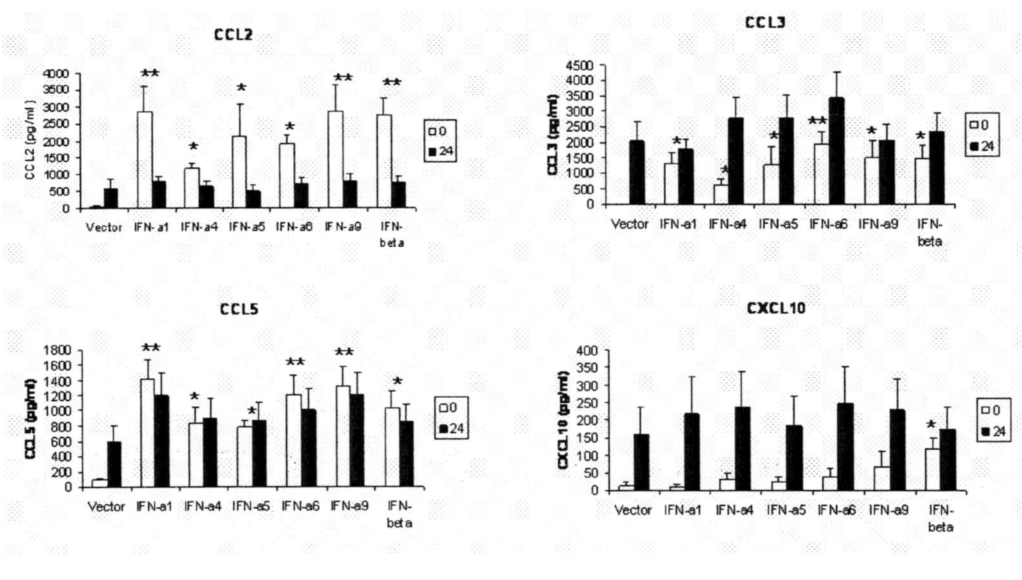

Figure 7. continued on next page.

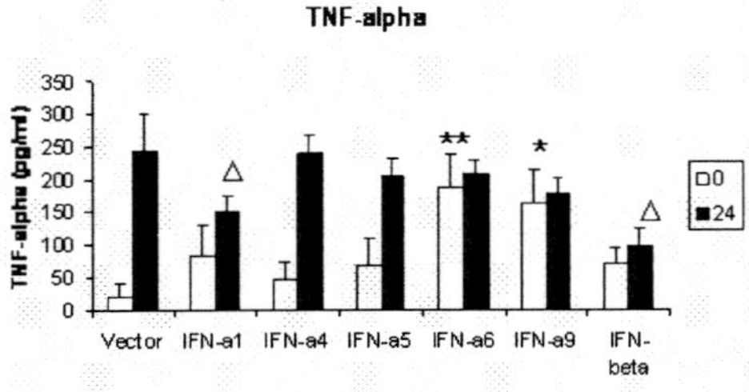

Figure 7. HSV-1 stimulates chemokine and TNF-α production by spleen cells. Spleen cells (1 x 10^6 cells/well) were treated with supernatants from vector or the indicated IFN transgene transfected L929 cells containing 100 units of biologically active IFN. The supernatant from non-transfected cells served as the control for chemokine levels and did not differ from vector supernatant exposed cells. Following a 24 hr incubation period, 200 μl of the supernatant was removed and assessed for chemokine/cytokine content, and the cells were infected with HSV-1 (MOI = 0.5). Twenty four hr post infection, the supernatant was removed and assayed for chemokine or cytokine (TNF-α) content by ELISA. This figure is a summary of 4-5 experiments. Bars represent the mean +/- SEM. **p<.01, *p<.05 comparing the IFN-treated to vector level for the respective chemokine or TNF-α. $^{\Delta}$p<.05 comparing IFN-α1 or IFN-β treated, HSV-1 infected spleen cells to the vector supernatant-treated, HSV-1-infected spleen cells.

CONCLUSION

In the ocular mouse model, HSV-1 is known to traffic to the TG via retrograde transport and infect neurons and satellite cells ultimately establishing a latent infection or trafficking further to the central nervous system and replicating uncontrollably resulting in the death of the animal. As indicated earlier, HSV-1 has a number of counter measures to block the host innate and adaptive immune response. Although production of type I IFN is noted early in the TG following viral infection of the eye [31], HSV-1 is capable of "colonizing" the ganglion except under cases in which the host over-expresses type I IFN [24]. Through the expression of viral proteins that specifically target the anti-viral IFN response [13-16] and thus suppress a potent, endogenous, anti-viral axis, the virus is capable of further replication in order to establish latency. The present findings offer yet another level of attenuating the host response by blocking chemokine production by TG cells in the presence of type I IFNs. Reducing the infiltration of leukocytes including macrophages and T lymphocytes into the TG during the initial stage of acute infection would ultimately lead to greater viral burden as demonstrated in mice deficient in T cells or macrophages [32, 33].

The association between HSV-1-induced chemokine suppression in type I IFN-treated TG cells and elevated SOCS-3 expression is consistent with IFN activation of the Jak/Stat pathway and the activation of the natural feedback inhibitor family, SOCS [34, 35]. Previous studies have reported the activation of the Jak/Stat pathway by chemokines including CCL2, CCL5, and CXCL12 [36-38] and in one report, SOCS-3 inactivation of a chemokine receptor, CXCR4 [39]. The present findings extend these observations implicating SOCS-3 in the HSV-1-mediated suppression of CCL2, CCL5, and CXCL10 production by IFN-stimulated primary TG cells. The mechanism by which this may occur in primary TG cells is currently unknown. However, in other cell systems, Stat-1 is implicated in the expression of CCL2,

CCL5, and CXCL10 [40-42] implying the suppression of Stat-1 activation (phosphorylation) by SOCS would suppress CCL2, CCL5, and CXCL10 expression.

As an alternative to SOCS-3 suppression of the Jak1/Stat-1 pathway, HSV-1-infected cells may undergo apoptosis or necrosis as a result of infection. However, we found little evidence to support apoptosis as a principal factor that would explain the significant reduction in chemokine levels following HSV-1 infection of TG cells compared to uninfected controls. Since HSV-1 infection typically culminates in a cytopathic effect of cells, it is possible that primary TG cells undergo significant cell loss by necrosis. However, this event would not explain the increase in CCL5 levels following HSV-1 infection of primary TG cells.

Although the list is not exhaustive, primary TG cells were found to constitutively secrete significant amounts of CCL2 as previously reported for HSV-1-latently infected TG cells. [31] However, CCL5 and CXCL10 were not detected or modest amounts were produced in the absence of stimulation. Following type I IFN stimulation, CCL2, CCL5, and CXCL10 levels were greatly enhanced or induced depending on the type I IFN species employed. However, CCL3 and TNF-α were never detected in the TG cells following IFN stimulation or HSV-1 infection. These results are in contrast to the *in vivo* situation in which CCL3 and TNF-α are detected in the TG of HSV-1-infected mice [6, 7, 25]. Taken together, these results suggest that CCL3 and TNF-α production in the TG following ocular HSV-1 infection is principally due to infiltrating leukocytes that are absent in the TG cell cultures of the present study. Consistent with this notion, spleen cells stimulated with type I IFN or infected with HSV-1 produced copious amounts of CCL3 and TNF-α. Unlike TG cells, IFN-stimulated spleen cells infected with HSV-1 showed no significant change in chemokine production with the exception of CCL2. Although macrophages are a likely source of CCL2 [43] and relatively resistant to HSV-1 infection, subpopulations of macrophages respond differently to the virus [44, 45]. Therefore, macrophages may also respond differently in chemokine production following HSV-1 infection. Likewise, additional hematopoietic- and non-hematopoietic-derived cells in the spleen cell preparations produce CCL2. For example, CCL2 is produced by murine fibroblasts [46] and both TG cell and spleen cell preparations would contain fibroblasts. Consequently, it is possible the majority of CCL2 is fibroblast-derived and infection of the fibroblasts with HSV-1 would likely lead to a reduction in CCL2 production as a result of viral-mediated fibroblast lysis.

Unlike the other type I IFNs, IFN-α1 and IFN-β suppressed HSV-1-mediated enhancement of TNF-α in spleen cell cultures. While the IFN-α1 effect on TNF-α production has not been previously reported, IFN-β is a known anti-inflammatory cytokine [47, 48] that suppresses LPS-induced TNF-α by monocytes [49]. Since TNF-α is a primary cytokine associated with pathological manifestations following ocular HSV-1 infection [50], the anti-viral, anti-inflammatory nature of IFN-β may be highly useful in preserving the visual axis [51].

Whereas type I IFNs operate through the same receptor [23], the present results consistently show variability in the action of selective IFN-αs and IFN-β in modifying chemokine production by TG or spleen cells or anti-viral efficacy in spleen cells. The diversity in chemokine induction by type I IFNs is also reflected in up-regulation of protein tyrosine kinases and the phosphorylation of Stat1 in TG cells. Another recent study has also described varying degrees of Jak/Stat activation as a result of type I IFN exposure [22]. These results are consistent with the range of biological outcomes elicited by type I IFNs and the response by different cell types to the IFNs [52, 53]. On a unit basis, IFN-β elicited the most robust response compared to the IFN-α in nearly every case. IFN-β has previously been reported to elicit higher levels of Jak1/Stat1 activation [54] and is thought to be an immediate early anti-viral response gene [55]. Therefore, IFN-β alone or in combination with other IFN-αs [26] or IFN-γ [56] may be a useful system to develop anti-viral therapies directed against ocular HSV-1 infection. The *in vitro* results using primary TG cells show no additive or synergistic action of type I IFNs (i.e., IFN-α1 and IFN-β) measuring HSV-1 replication. However, these same experiments did find an additive effect on selective chemokines including CCL2 and CCL5 even in the presence of HSV-1. The consequences of this finding *in vivo* have yet to be explored.

ACKNOWLEDGMENTS

The authors are grateful to Drs. Cassie M. James and Vanessa Cull for the type I IFN plasmid constructs. The authors would also like to thank Mrs. Benitta John-Philip for her technical support. This work was supported by USPHS grant AI53108, an unrestricted grant from Research to Prevent Blindness (RPB), and an RPB Stein Research Professorship award.

REFERENCES

[1] Shimeld, C; Whiteland, JL; Nicholls, SM; Grinfeld, E; Easty, DL; Gao, H; Hill, TJ. (1995). Immune cell infiltration and persistence in the mouse trigeminal ganglion after infection of the cornea with herpes simplex virus type 1. *J. Neuroimmunol.* **61**, 7-16.

[2] Liu, T; Tang, Q; Hendricks, RL. (1996). Inflammatory infiltration of the trigeminal ganglion after herpes simplex virus type 1 corneal infection. *J. Virol.* **70**, 264-271.

[3] Thomas, J; Gangappa, S; Kanagat, S; Rouse, BT. (1997). On the essential involvement of neutrophils in the immunopathologic disease. Herpetic Stromal Keratitis. *J. Immunol.* **158**, 1383-1391.

[4] Halford, WP; Gebhardt, BM; Carr, DJJ. (1996). Persistent cytokine expression in trigeminal ganglion latently infected with herpes simplex virus type 1. *J. Immunol.* **157**, 3542-3549.

[5] Su, YS; Yan, XT; Oakes, JE; Lausch, RN. (1996). Protective antibody therapy is associated with reduced chemokine transcripts in herpes simplex virus type 1 corneal infection. *J. Virol.* **70**, 1277-1281.

[6] Tumpey, TM; Cheng, H; Yan, XT; Oakes, JE; Lausch, RN. (1998). Chemokine synthesis in the HSV-1-infected cornea and its suppression by interleukin-10. *J. Leukoc. Biol.* **63**, 486-492.

[7] Tumpey, TM; Cheng, H; Cook, DN; Smithies, O; Oakes, JE; Lausch, RN. (1998). Absence of macrophage inflammatory protein-1α prevents the development of blinding herpes stromal keratitis. *J. Virol.* **72**, 3705-3710.

[8] Carr, DJJ; Chodosh, J; Ash, J; Lane, TE. (2003). Effect of anti-CXCL10 monoclonal antibody on herpes simplex virus type 1 keratitis and retinal infection. *J. Virol.* **77**, 10037-10046.

[9] Jennings, SR; Rice, PL; Kloszewski, ED; Anderson, RW; Thompson, DL; Tevethis, SS. (1985). Effect of herpes simplex virus type 1 and 2 on surface expression of class I major histocompatability complex antigens on infected cells. *J. Virol.* **56**, 757-766.

[10] Hill, A; Jugovic, P; York, I; Russ, G; Bennink, J; Yewdell, J; Ploegh, HH; Johnson, D. (1995). Herpes simplex virus turns off the TAP to evade host immunity. *Nature* **375**, 411-415.

[11] Neumann, J; Eis-Hubinger, AM; Koch, N. (2003). Herpes simplex virus type 1 targets the MHC class II processing pathway for immune evasion. *J. Immunol.* **171**, 3075-3083.

[12] Nagashunmugam, T; Lubinski, J; Wang, L; Goldstein, LT; Weeks, BS; Sundaresan, P; Kang, EH; Dubin, G; Friedman, HM. (1998). In vivo immune evasion mediated by the herpes simplex virus type 1 immunoglobulin G Fc receptor. *J. Virol.* **72**, 5351-5359.

[13] Chou, J; Chen, JJ; Gross, M; Roizman, B. (1995). Association of a Mr 90,000 phosphoprotein with protein kinase PKR in cells exhibiting enhanced phosphorylation of translation initiation factor eIF-2α and premature shutoff of protein synthesis after infection with $\gamma_1 34.5^-$ mutants of herpes simplex virus 1. *Proc. Natl. Acad. Sci. USA* **92**, 10516-10520.

[14] Mossman, KL; Saffran, HA; Smiley, JR. (2000). Herpes simplex virus ICPO mutants are hypersensitive to interferon. *J. Virol.* **74**, 2052-2056.

[15] Harle, P; Sainz, B Jr; Carr, DJJ; Halford, WP. (2002). The immediate-early protein, ICPO, is essential for the resistance of herpes simplex virus to interferon-α/β. *Virology* **293**, 295-304.

[16] Peters, GA; Khoo, D; Mohr, I; Sen, GC. (2002). Inhibition of PACT-mediated activation of PKR by the herpes simplex virus type 1 Us11 protein. *J. Virol.* **76**, 11054-11064.

[17] Yokota, SI; Yokosawa, N; Kubota, T; Suzutani, T; Yoshida, I; Miura, S; Jimbow, K; Fujii, N. (2001). Herpes simplex virus type 1 suppressess the interferon signaling pathway by inhibiting phosphorylation of STATs and janus kinases during an early infection stage. *Virology* **286**, 119-124.

[18] Tortorella, D; Gewurz, BE; Furman, MH; Schust, DJ; Ploegh, HL. (2000). Viral subversion of the immune system. *Annu. Rev. Immunol.* **18**, 861-926.

[19] Liesegang, TJ. (1999). Classification of herpes simplex virus keratitis and anterior uveitis. *Cornea* **18**, 127-143.

[20] Sawtell, NM. (1998). The probability of in vivo reactivation of herpes simplex virus type 1 increases with the number of latently infected neurons in the ganglia. *J. Virol.* **72**, 6888-6892.

[21] Carr, DJJ; Al-khatib, K; James, CM; Silverman, R. (2003). Interferon-β suppresses herpes simplex virus type 1 replication in trigeminal ganglion cells through an RNase L-dependent fashion. *J. Neuroimmunol.* 141, 40-46.

[22] Cull, VS; Tilbrook, PA; Bartlett, EJ; Brekalo, NL; James, CM. (2003). Type I interferon differential therapy for erythroleukemia: specificity of STAT activation. *Blood* **101**, 2727-2735.

[23] Stark, GR; Kerr, IM; Williams, BRG; Silverman, RH; Schreiber, RD. (1998). How cells respond to interferons. *Annu. Rev. Biochem.* **67**, 227-264.

[24] Carr, DJJ; Veress, LA; Noisakran, S; Campbell, IL. (1998). Astrocyte-targeted expression of IFN-α1 protects mice from acute ocular herpes simplex virus type 1 infection. *J. Immunol.* **161**, 4859-4865.

[25] Shimeld, C; Whiteland, JL; Williams, NA; Easty, DL; Hill, TJ. (1997). Cytokine production in the nervous system of mice during acute and latent infection with herpes simplex virus type 1. *J. General Virology* **78**, 3317-3325.

[26] Bartlett, EJ; Cull, VS; Brekalo, NL; Lenzo, JC; James, CM. (2002). Synergy of type I interferon-A6 and interferon-B naked DNA immunotherapy for cytomegalovirus infection. *Immunol. Cell Biology* **80**, 425-435.

[27] Ito, M; Watanabe, M; Kamiya, H; Sakurai, M. (1997). Herpes simplex virus type 1 induces apoptosis in peripheral blood T lymphocytes. *J. Infect. Dis.* **175**, 1220-1224.

[28] Raftery, MJ; Behrens, CK; Muller, A; Krammer, PH; Walczak, H; Schonrich, G. (1999). Herpes simplex virus type 1 infection of activated cytotoxic T cells: Induction of fratricide as a mechanism of viral immune evasion. *J. Exp. Med.* **190**, 1103-1114.

[29] Yasukawa, H; Sasaki, A; Yoshimura, A. (2000). Negative regulation of cytokine signaling pathways. *Annu. Rev. Immunol.* **18**, 143-164.

[30] Yokota, SI; Yokosawa, N; Okabayashi, T; Suzutani, T; Miura, S; Jimbow, K; Fujii, N. (2004). Induction of suppressor of cytokine signaling-3 by herpes simplex virus type 1 contributes to inhibition of the interferon signaling pathway. *J. Virol.* **78**, 6282-6286.

[31] Carr, DJJ; Noisakran, S; Halford, WP; Lukacs, N; Asensio, V; Campbell, IL. (1998). Cytokine and chemokine production in HSV-1 latently infected trigeminal ganglion cell cultures: Effects of hyperthermic stress. *J. Neuroimmunol.* **85**, 111-121.

[32] Cheng, H; Tumpey, TM; Staats, HF; Van Rooijen, N; Oakes, JE; Lausch, RN. (2000). Role of macrophages in restricting herpes simplex virus type 1 growth after ocular infection. *Invest. Ophthalmol. Vis. Sci.* **41**, 1402-1409.

[33] Carr, DJJ; Noisakran, S. (2002). The antiviral efficacy of the murine alpha-1 interferon transgene against ocular herpes simplex virus type 1 requires the presence of $CD4^+$, α/β T-cell receptor-positive T lymphocytes with the capacity to produce gamma interferon. *J. Virol.* **76**, 9398-9406.

[34] Endo, TA; Masuhara, M; Yokouchi, M; Suzuki, R; Sakamoto, H; Mitsui, K; Matsumoto, A; Tanimura, S; Ohtsubo, M; Misawa. (1997). A new protein containing an SH2 domain that inhibits JAK kinases. *Nature* **387**, 917-921.

[35] Naka, T; Narazaki, M; Hirata, M; Matsumoto, T; Minamoto, S; Aano, A; Nishimoto, N; Kajita, T; Taga, T; Yoshizaki, K. (1997). Structure and function of a new STAT-induced STAT inhibitor. *Nature* **387**, 924-929.

[36] Vila-Coro, AJ; Rodriguez-Frade, JM; Martin De Ana, AM; Moreno-Ortiz, MC; Martinez, C; Mellado, M. (1999). The chemokine SDF-1α triggers CXCR4 receptor dimerization and activates the Jak/Stat pathway. *FASEB J.* **13**, 1699-1710.

[37] Rodriguez-Frade, JM; Vila-Coro, A.; Martin De Ana, AM; Albar, JP; Martinez, C; Mellado, M. (1999). The chemokine monocyte chemoattractant protein-1 induces functional responses through dimerization of its receptor CCR2. *Proc. Natl. Acad. Sci. USA* **96**, 3628-3633.

[38] Rodriguez-Frade, JM; Vila-Coro, AJ; Martin De Ana, AM; Nieto, M; Sanchez-Madrid, F; Proudfoot, AEI; Wells, TNC; Martinez, C; Mellado, M. (1999). Similarities and

differences in RANTES- and (AOP)-RANTES-triggered signals: implications for chemotaxis. *J. Cell Biol.* **144**, 755-765.

[39] Soriano, SF; Hernanz-Falcon, P; Rodriguez-Frade, JM; Martin De Ana, A; Garzon, R; Carvalho-Pinto, C; Vila-Coro, AJ; Zaballos, A; Balomenos, D; Martinez, C; Mellado, M. (2002). Functional inactivation of CXC chemokine receptor 4-mediated responses through SOCS-3 up-regulation. *J. Exp. Med.* **196**, 311-321.

[40] Ohmori, Y; Hamilton, TA. (1995). The interferon-stimulated response element and a kappa B site mediate synergistic induction of murine IP-10 gene transcription by IFN-gamma and TNF-alpha. *J. Immunol.* **154**, 5235-5244.

[41] Cremer, I; Ghysdael, J; Viellard, V. (2002). A non-classical ISRE/ISGF3 pathway mediates induction of RANTES gene transcription by type I IFNs. *FEBS. Lett.* **511**, 41-45.

[42] Ikeda, T; Sato, K; Kuwada, N; Matsumura, T; Yamashita, T; Kimura, F; Hatake, K; Ikeda, K; Motoyoshi, K. (2002). Interleukin-10 differently regulates monocyte chemoattractant protein-1 gene expression depending on the environment in a human monoblastic cell line, UG3. *J. Leukoc. Biol.* **72**, 1198-1205.

[43] Frazier-Jessen, MR; Kovacs, EJ. (1995). Estrogen modulation of JE/monocyte chemoattractant protein-1 mRNA expression in murine macrophages. *J. Immunol.* **154**, 1838-1845.

[44] Leary, K; Connor, JR; Morahan, PS. (1985). Comparison of herpes simplex virus type 1 DNA replication and virus production in murine bone marrow-derived and resident peritoneal macrophages. *J. General Virology* **66**, 1123-1129.

[45] Morahan, PS; Mama, S; Anaraki, F; Leary, K. (1989). Molecular localization of abortive infection of resident peritoneal macrophages by herpes simplex virus type 1. *J. Virol.* **63**, 2300-2307.

[46] Van-Damme, J; Decock, B; Bertini, R; Conings, R; Lenaerts, JP; Put, W; Opdenakker, G; Mantovani, A. (1991). Production and identification of natural monocyte chemotactic protein from virally infected murine fibroblasts. Relationship with the product of the mouse competence (JE) gene. *Eur. J. Biochem.* **199**, 223-229.

[47] Khademi, M; Wallstrom, E; Andersson, M; Piehl, F; Di Marco, R; Olsson, T. (2000). Reduction of both pro- and anti-inflammatory cytokines after 6 months of interferon beta-1a treatment of multiple sclerosis. *J. Neuroimmunol.* **103**, 202-210.

[48] Muraro, PA; Liberati, L; Bonanni, L; Pantalone, A; Caporale, CM; Iarlori, C; De Luca, G; Farina, D; Lugaresi, A; Gambi, D. (2004). Decreased integrin gene expression in patients with MS responding to interferon-β treatment. *J. Neuroimmunol.* **150**, 123-131.

[49] Molnarfi, N; Gruaz, L; Dayer, JM; Burger, D. (2004). Opposite effects of IFN beta on cytokine homeostasis in LPS- an cell contact-activated human monocytes. *J. Neuroimmunol.* **146**, 76-83.

[50] Wasmuth, S; Bauer, D; Yang, Y; Steuhl, KP; Heiligenhaus, A. (2003). Topical treatment with antisense oligonucleotides targeting tumor necrosis factor-α in herpetic stromal keratitis. *Invest. Ophthalmol. Vis. Sci.* **44**, 5228-5234.

[51] Minagawa, H; Sakai, Y; Li, YY; Ishibashi, T; Inomata, H; Mori, R. (1997). Suppression of infectious virus spread and corneal opacification by the combined use of recombinant interferon beta and interleukin-10 following corneal infection with herpes simplex virus-1 in mice. *Antiviral Res.* **36**, 99-105.

[52] Breirley, MM; Fish, EN. (2002). IFN-α/β receptor interactions to biologic outcomes: Understanding the circuitry. *J. Interferon Cytokine Res.* **22**, 835-845.

[53] Gomez, D; Reich, NC. (2003). Stimulation of primary human endothelial cell proliferation by IFN. *J. Immunol.* **170**, 5373-5381.

[54] Grumbach, IM; Fish, EN; Uddin, S; Majchrzak, B; Colamonici, OR; Figulla, HR; Heim, A; Platanias, LC. (1999). Activation of the Jak-Stat pathway in cells that exhibit selective sensitivity to the antiviral effects of IFN-β compared with IFN-α. *J. Interferon Cytokine Res.* **19**, 797-801.

[55] Marie, I; Durbin, JE; Levy, DE. (1998). Differential viral induction of distinct interferon-α genes by positive feedback through interferon regulatory factor-7. *EMBO J.* **17**, 6660-6669.

[56] Sainz, B Jr; Halford, WP. (2002). Alpha/beta interferon and gamma interferon synergize to inhibit the replication of herpes simplex virus type 1. *J. Virol.* **76**, 11541-11550.

In: Chemokine Research Trends
Editor: Leah R. Grinwald, pp. 47-60

ISBN: 978-60021-588-9

Chapter 3

THE ROLE OF CHEMOKINES IN THE PATHOGENESIS OF HIV-1 DURING HIV-1/TB DUAL INFECTION

Zahra Toossi*

Division of Infectious Disease, Department of Medicine,Case Western Reserve University, Cleveland, OH 44106-4984, 4, Veterans' Administration Medical Center Cleveland OH 44106

ABSTRACT

Worldwide, co-infection by Mycobacterium tuberculosis is common among HIV-1-infected subjects. These two pathogens influence the prevalence, morbidity, and mortality caused by one another. Tuberculosis (TB) occurs earlier during HIV-1 disease (CD4 cell count over 200/μl) in two thirds of dually infected subjects, and is associated with accelerated rates of HIV-1 disease progression. Higher viral loads, increased HIV-1 diversity, and changes in cytokine and chemokine levels at sites of M. tuberculosis infection (pleural or pulmonary TB) have been well-documented in dually infected subjects and likely all contribute to the negative impact of TB on HIV disease. Increased expression of CCR5 in combination with relatively low levels of HIV-1-inhibitory β chemokines is conducive to HIV-infection of lymphocytes and macrophages that have recently migrated to M. tuberculosis infected organs. Together with accentuated HIV-1 replication, which is guaranteed by excesses in TNF-α and MCP-1, a state of saturation of infection by CCR5-tropic HIV in *situ* results. Due to excessive cellular activation, however, preferential deletion of these HIV-infected cells may occur. On the other hand, increased expression of CXCR4 by macrophages supported by excess TGF-β1 in situ supports enhanced infection by CXCR-4 tropic HIV-1. This may allow expansion of the more cytopathic viral strains in *situ*, and ultimately contribute to a predominance by CXCR-4 tropic HIV-1 infection. Increased levels of IFN-γ and TNF-α in *situ* guarantee cellular activation, which is necessary for successful infection through either chemokine receptor. This review addresses the role of chemokines and their receptors at sites of active tuberculosis in progression of HIV-1 infection in dually infected subjects.

* E-mail address: zxt2@po.cwru.edu, Phone: 216-368-4844, FAX 216-368-2034, Correspondence: Zahra Toossi, M.D., Division of Infectious Disease, Department of Medicine, Case Western Reserve University, 11100 Euclid Ave., Cleveland, OH 44106

INTRODUCTION

Progress in understanding the basis of human immunodeficiencey virus 1 (HIV-1) infection has identified the seminal role that hiv-1 co-receptorS (namely CCR5 and CXCR4) expression and usage play in the pathogenesis of HIV-1 disease. Understanding how chemokines and cytokines and their receptors are altered by co-infections of HIV-1 disease, such as TB, is critical in the design and application of reagents that can effectively inhibit progression of HIV-1 disease in co-infected individuals. Wheras understanding of the underlying reasons for transition of CCR5 tropic isolates, that are predominant during early years of HIV-1 disease, TO CXCR4 tropic HIV-1 isolates which are more cytopathic is ongoing, studies identifying this process at sites of TB may be incitful of viral pathogenesis during co-infections. Recent research has identified potent inhibitors of HIV-1 infection based on the structure of chemokine receptors, which may be therapeutic options in management of dual HIV-1/TB.

Here we will first review recent findings on the role of chemokine receptors in pathogenesis of HIV-1 infection. Several good review articles have been published more recently in this regard [1, 2], and the reader is referred to these for a better appreciation of the role of chemokine receptors in transmission and pathogenesis of HIV-1 disease. These understandings have been applied in part to dual HIV/TB infection, that remains the most common co-infection of HIV-1 disease. As HIV-1/TB dual infection is associated with progression of HIV-1 disease and is associated with significant mortality and morbidity worldwide, findings in this model may prove to be important in further development of immunoadjuvants. Evidence for replication of both CCR5 and CXCR4 tropic HIV-1 at sites of active MTB infection during TB is accumulating. This appears to be based on the cytokine/chemokine milieu associated with human TB in situ, and is likely to be related to the chronicity of MTB infection. Whether the now well-documented progression of HIV-1 disease in subjects with dual HIV-1/TB is related to expansion of viral phenotype to include both CCR5 and CXCR4, however, remains to be clarified.

The Role of Chemokine and Chemokine Receptors in Pathogenesis of HIV-1 Infection

Entry of HIV-1 into target cells involves interactions of the viral envelope protein, gp120, with CD4 and a co-receptor, which is predominantly one of two G-coupled transmembrane chemokine receptors, namely CCR5 or CXCR4. Subsequent to binding of viral envelope to host cell CD4 and either CCR5 or CXCR4, several essential conformational rearrangements in the viral envelope occur. This results in creation of structural intermediates of HIV-1 and ultimately exposure of viral gp41, which mediates fusion of viral and cellular membranes [3]. The central role of CCR5 and CXCR4 in pathogenesis of HIV-1 disease and the pattern of co-receptor usage by HIV-1 during the spectrum of disease has become more clear in recent studies. While a dozen other chemokine receptors have been shown to mediate HIV-1 entry in cell lines *in vitro* [4], their clinical relevance remains undefined. Importantly, there is no correlation between HIV-1 disease progression and the use of co-receptors other than with CCR5 and CXCR4 [5, 6].

The importance of chemokine co-receptors in pathogenesis of HIV-1 infection is underscored by extremely low rates of infection in subjects with genetic mutation in CCR5 gene (△32-CCR5 mutation) [7, 8]. Whereas, on rare occasion, some of these individuals may become infected with CXCR4 tropic viral isolates [9, 10], they appear to be protected from infection through other chemokine receptors. Further, the progression of HIV-1 disease is reduced by heterozygocity for △32-CCR5 mutation among infected individuals [11].

Studies have shown that the selective replication and dominance of CCR5 tropic HIV-1 in early years after infection is due to cellular environments in which HIV-1 enters to establish a successful infection and in which it proliferates. At both gastro-intestinal and genital sites of HIV-1 infection, CCR5 tropic virus is promoted almost exclusively due to co-expression of CCR5 on CD4 positive dendritic cells (DC), macrophages and T-cells. On the other hand the expression of CXCR4 on these cellular targets of HIV-1 at sites of infection appears to be more limited. For example, a 10 fold higher expression of CCR5 as compared to CXCR4 has been documented in the macaque model on vaginal epithelial DC [12]. The gut associated lymphoid tissue (GALT) also has low expression of CXCR4 [13]. Further, there is also some evidence that X4 HIV-1 infection may be actually blocked at sites of infection because of excessive levels of SDF-1, the natural ligand for CXCR4 [14]. Importantly cellular activation, which is absolutely necessary to the success of HIV-1 infection of CD4 + T-cells [15], is readily provided at gastrointestinal and genital sites of infection that are continuously exposed to micro-organisms.

In general, CCR5 using HIV-1 strains do not induce syncytium (NSI), and replicate more slowly (compared to CXCR4 HIV-1 that induce syncytia [SI]), and while they predominate during early disease, they can be isolated at all stages of disease. Phenotypic switch from CCR5 using/NSI to CXCR4 using/SI is believed to be an in vivo characteristic of the virus subjected to immunologic pressure. Interestingly, the switch from CCR5 using to CXCR4 using virus involves mutation in 2-3 amino acids in gp120 only [16, 17]. In approximately 50% of HIV-1 infected subjects after about 5 years of infection, CXCR4 using SI viruses become detectable [18]. Whereas, phenotypic switch is well known to be associated with poor prognosis [19], development of AIDS in its absence is common. In macaques, CCR5 tropic SIV evolves to become more virulent but without broadening of tropism for CXCR4 usage [20]. The same phenomenon has also been documented for HIV-1 in humans [21]. In HIV-1 infected subjects, the emergence of CXCR4 using HIV-1 variants has been documented to occur around the time of early to intermediate phase of infection, peak at the end of the intermediate stage and decline during late phases of disease [22]. Clonal isolates of HIV-1 that are capable to use both CCR5 and CXCR4 co-receptors (i.e.true R5X4) [23, 24] have been shown in the vitro. However, it is likely that most clinical isolates that grow on both CCR5- and CXCR4- positive cells represent a mixture of R5 and X4, rather than being truly dual tropic [1, 2]. This scenario may have in vivo relevance in HIV-1 disease during opportunistic infections (OI). In the absence of a clear co-receptor change from NSI to SI, increased cytopathicity of R5 correlates with a progressive course of disease [25]. Clinical HIV-1 isolates isolated from patients gain replicative fitness over time, and X4 isolates can be demonstrated to have replicative advantage as compared to CCR5 isolates [26].

In the absence of OI, progression of HIV-1 disease is characterized by dysregulations in expression of β-chemokines and CCR5 that correlate with the level of CD4 immunodeficiencey [27]. These dysregulations appear to be compartmentalized to tissues, such as the lung. However, whether these underlie HIV-1 associated pathologies or predispose to development of OI is not clear. For example, bronchoalveolar lavage cells from HIV-infected subjects with low CD4 T-cell counts (<200/ μl), but not high CD4 T-cell counts, displayed limitations in stimulated release of MIP-1α [27]. Stimulated expression of MIP-1α was also lower by intracellular staining in blood mononuclear cells of HIV-1 infected subjects as compared to healthy individuals. Whereas stimulated expression of CCR5 was also lower in BAC from HIV-infected subjects, this limitation was not as extensive as that of MIP-1α expression. This discrepancy in expression of CCR5 and its major ligand (MIP-1α) may undelie expansion of HIV-1 proliferation and its associated pathology [28] in the lung.

Impact of TB on HIV-1 Disease

Worldwide, one in four HIV-1-infected subjects is diagnosed with active tuberculosis (TB) [[29], making it the most frequent opportunistic infection during HIV-1 disease. Progression of HIV-1 disease is faster in HIV-1/TB dually infected patients compared to HIV-1 infected subjects without TB [30, 31]. Whereas TB occurs throughout the course of HIV-1 disease, higher CD4 T-cell counts (300-400 cells/ μl) have been documented in over 2/3 of dually infected subjects [32]. The impact of TB on HIV-1 disease is highlighted by a recent prospective study; increased HIV-1 mortality was seen in HIV-1/TB dually infected subjects with higher CD4 T-cell counts (CD4 > 200/μl) in comparison to CD4-matched HIV-1 infected subjects [33]. Development of active TB is associated with both a systemic increase in HIV-1 load and expansion of viral genetic diversity [34, 35]. The increase in viral replication and heterogeneity is particularly intense at sites of Mycobacterium tuberculosis (MTB) infection [36] (see below). Mechanisms of increased viral activity in HIV-1/TB include both transcriptional activation of HIV-1 in infected cells and dissemination of viral infection to uninfected cells [37]. However, how these factors are linked to disease progression is still not clear. Understanding viral dynamics at sites of MTB infection may be conducive to development and application of therapies that control augmentation in HIV-1 activity, and may be useful in control of progression of HIV-1 disease in dual HIV-1/TB infection. Virologic approaches to HIV-1/TB is critical, now that it is clear that as opposed to other HIV-1 associated OI, TB leads to persistent HIV-1 replication which does not abate with anti-tuberculous therapy alone [38].

The capacity of MTB to induce production of β chemokines with known chemotactic activity for monocytes and lymphocytes has been investigated before [39]. Interestingly, while MTB stimulated the production of MCP-1 and MIP-1α in blood monocytes and alveolar macrophages of healthy subjects, no RANTES was induced. Furthermore, production of these chemokines was higher in cultures of mature alveolar macrophages than their precursors, blood monocytes. Both infection by avirulent and virulent MTB and stimulation by purified protein derivative of H37Rv (PPD), but not non-specific stimulation through phagocytosis of latex beads, induced production of β chemokines. Further, bronchoalveolar lavage fluid from patients with active pulmonary tuberculosis contained increased levels of RANTES, MCP-1, and IL-8 as compared with healthy control subjects [39]. Other studies have shown that MTB induces CCR5 mRNA expression in mononuclear cells *in vitro* [40].

In another study, MTB stimulated β chemokines were measured in blood mononuclear cells from HIV-1 infected and un-infected subjects with pulmonary TB and in HIV-1 infected and un-infected control individuals [41]. MTB stimulation of blood mononuclear cell models from TB patients model events likely to occur at sites of TB where cells are exposed to both pathogens. MTB-induced MIP–1α and CCR5 expression were lower in unseparated blood mononuclear cells of dually infected HIV-1/TB patients than in those of TB patients. MIP-1α was also lower in HIV-1 infected control patients than in those of healthy control subjects [41]. Defects in pro-inflammatory cytokines (TNF-α) were also observed. Whether defects in MIP-1α and/or other inflammatory chemokines underlies the predisposition to development of active TB during HIV-1 disease is still not clear. The induction of expression of mRNA for CCR5 by MTB in mononuclear cells from HIV-1/TB patients was higher than that of HIV-1 infected subjects without TB [41]. Whether this relates to augmentations of HIV-1 replication during TB is not clear.

Increased levels of CCR5 tropic HIV-1 (and not CXCR4 tropic virus) has been isolated from plasma of HIV-1/TB subjects with pulmonary TB only [42]. This likely is in part reflective of the earlier stage of HIV-1 disease at which TB occurs, and implies an intense immunologic pressure on CCR5 tropic HIV-1. However, this has to be understood in the context of the assays currently used to assess viral phenotype which are based on genotypic analysis of a limited numbers (10^3) of HIV-1 clones [36]. In addition, recent research indicates that the magnitude of virions released from lysis of CCR5 infected mononuclear cells (CCR5 burst size) is an order of magnitude greater than that of CXCR4 infected cells [43]. Therefore small numbers of CXCR4 tropic isolates may remain unidentified in plasma of dually infected HIV-1/TB subjects. More recently CXCR4 tropic HIV-1 has been isolated from alveolar macrophages of HIV-1/TB patients with pulmonary TB [44]. Of note, we have not isolated CXCR4 tropic isolates in pleural TB [45]. Whether this implies a predominant effect of site of active TB (and its cytokine and chemokine milieu) on replication and selection of HIV-1 isolates is not clear. In addition, isolation of CCR5 vs CXCR4 tropic HIV-1 may be dependent on length of time after which TB is diagnosed which is reflective of the duration of immunologic pressure by MTB on the virus.

Chemokine and Chemokine Receptor Characteristics of Sites of TB in HIV-1/TB Dual Infection

Recent studies have shown that HIV-1 replication at sites of active MTB infection in HIV-1/TB patients is substantial [45, 46]. An associated increase in HIV-1 heterogeneity in virus isolated from sites of MTB infection has been documented [36]. In the pleural TB, which is a common presentation of TB in dually infected HIV-1/TB patients in Africa [47], HIV-1 replication is augmented both in the cellular and acellular compartments of the pleural fluid [45]. Mechanisms for HIV-1 expansion at sites of TB include both transcriptional activation [48] and predisposition to new rounds of viral infection of mononuclear cells recruited to sites of MTB infection. These mechanisms may account for enhanced viral activity and dissemination of HIV-1 during TB.

Levels of TNF–α, a macrophage activating cytokine that supports HIV-1 replication strongly [49, 50], is augmented during TB [51] and at sites of MTB infection [52]. Excess expression of TNF-α at sites of MTB infection is supported by the predominant TH1 cytokine profile at sites of MTB infection represented by excess IFN-γ [52]. IFN-γ supports expression of CCR5 on all mononuclear cells [53], and thus supports expression of CCR5 tropic HIV-1 at sites of TB. In fact, CCR5 mRNA was increased in mononuclear cells from pleural sites of

MTB infection in HIV-1/TB patients [45]. Together with a relative paucity of HIV-1 inhibiting β-chemokines in pleural fluid, the environment is conducive to enhanced infection of newly recruited mononuclear cells by CCR5 tropic HIV-1. Mechanistically, TNF−α activates HIV-1 LTR driven transcription [54], and neutralization of TNF−α inhibited in part the induction of HIV-1 p24 production in monocytes and macrophages infected by HIV-1 *in vitro* [48, 55]. TNF−α is up-regulated during active TB [51], at sites of MTB infection [56], and is induced by MTB stimulation of mononuclear cells *in vitro* [57]. Concentrations of TNF−α were significantly higher in the pleural fluid as compared to plasma of HIV-1 infected patients with pleural TB [45]. Neutralization of TNF−α partially abrogated the increased viral replication of HIV-1 infected mononuclear cells which were co-cultured with pleural fluid from TB patients [56]. Whereas early inhibition of TNF−α during HIV-1/TB may be helpful in down-modulation of HIV-1 replication, however, due to delays in diagnosis of TB, use of TNF−α inhibitors as an immunotherapeutic measure has not been convincing to date [58, 59].

The limitation in HIV-1 inhibitory molecules at sites of MTB infection [45] involves the CCR5-using β-chemokines alone. Levels of another β-chemokine, monocyte chemoattractant protein (MCP)-1 which binds to CCR2 (instead of CCR5) [60], and does not block HIV-1 entry, was expanded in both pleural and pulmonary TB [39, 45]. MCP-1 is involved in the recruitment of monocytes and memory T-cells to MTB-infected tissues [61], which may then be targeted for new rounds of HIV-1 infection at sites of MTB infection. Thus, in face of limited β-chemokines that are inhibitory to HIV-1 and augmented CCR5 expression in situ , new rounds of HIV-1 infection appear particularly likely (Figure 1).

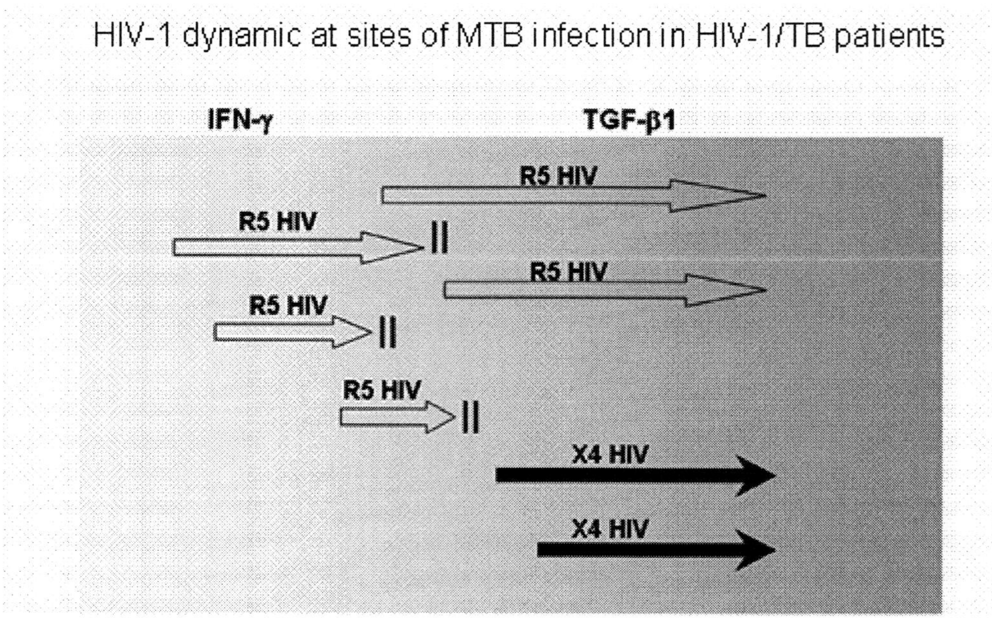

Figure 1. Replication of CCR5(R5) and CXCR4(X4) HIV-1 is supported by IFN-γ and TGF-β1 at sites of TB. Inhibition of R5 HIV-1(II) by TGF-β1 is conductive to further expansion of HIV-1 at sites of MTB infection.

Due to continuous recruitment of blood monocytes to sites of MTB infection, active pulmonary TB is associated with an immature macrophage alveolitis [62]. Monocytes and immature macrophages have an increased susceptibility to HIV-1 infection [63], and have been shown to be the major source of HIV-1 during OI [64]. Interestingly differentiation of monocytes to macrophages *in vitro* is associated with a reversal of the effect of MTB from one that allows active HIV-1 replication to viral suppression [65]. HIV-1 infected macrophages may in fact serve as reservoirs of HIV-1 throughout the course of HIV-1 disease. Recent studies indicate that macrophages are a significant source of HIV-1 during active pulmonary TB [66]. CXCR4 tropic HIV-1 has been found to originate predominantly from macrophages from pulmonary sites of MTB infection [44]. Cytokines abundant at sites of MTB infection (i.e TGF-β1 and IFN-γ) [52, 67], may have a significant role in promotion of CXCR4 tropic HIV-1 infection in macrophages (Toossi Z, submitted). TGF-β1 induced expression of CXCR4 at both gene and protein levels in macrophages derived from blood monocytes (MDM) in vitro, but not in mature autologous alveolar macrophages. TGF-β1 supported HIV-1 infection of MDM by a clinical strain of CXCR4 tropic HIV-1, which was inhibited significantly by AMD 3100 [3] (inhibitor of CXCR4 tropic HIV-1 infection). IFN-γ, which inhibits the majority of biological effects of TGF-β1 [68], was synergistic with TGF-β1 in HIV-1 infection of MDM by CXCR4 tropic HIV-1. Expression of CXCR4 on pleural macrophages of majority of HIV-1/TB subjects with pleural TB was increased over that in autologous blood monocytes. Further, pleural fluid from TB patients (rich both in TGF-β1 and IFN-γ) supported infection of MDM from healthy subjects by CXCR4 tropic HIV-1. Therefore at sites of TB, TGF-β1 is supportive of CXCR4 tropic HIV-1 infection (Figure 1). On the other hand, evidence suggests that TGF-β1 may inhibit infection by CCR-5 tropic HIV-1[50] . Theferfore, the milieu at sites of TB may be conducive to a co-receptor usage switch from CCR5 to CXCR4 tropic HIV-1 infection.

Therapeutic Considerations

The intense research on identification of chemokine co-receptors that mediate HIV-1 entry has lead to identification of several reagents that can be used as anti-viral therapies reviewed in [3, 69]. Whereas reagents that target either CCR5 or CXCR4 await full clinical trials, the fusion inhibitors that target HIV-1 gp41 [70] and inhibit HIV-1 entry through either co-receptor, are already in clinical use [71-73].

Modifications in the N-terminus region of RANTES has lead to identification of several CCR5 antagonists [74], some of which may have clinical relevance in the future. Recent studies have in fact identified a role for CCR5 inhibitors in prevention of SHIV infection in the macaque model of HIV-1 infection [75]. The anti-HIV activity of these analogues appears to include both internalization and sequestration of CCR5 from the cell surface [76] as well as competitive binding with HIV-1 for CCR5 [77]. However, preclinical studies of the utility of RANTES analogues in vivo is limited [78, 79]. Further, at least one *in vitro* study indicates that their use may be problematic; AOP-RANTES can actually stimulate replication of primary SI/X4 HIV-1 isolates, as well as promote a breakthrough from its inhibitory activity in PBMC infected with primary NSI/R5 isolates [80]. Further, RANTES analogues other than PSC-RANTES were found to be digested by Cathepsin released from monocytes and neutrophils in vitro [81]. Interestingly, cervicovaginal washing from macaques was able to digest RANTES analogs also.

The CXCR4 antagonists, AMD3100 and KRH-1636, have shown potent anti-HIV-1 activity in vitro [3]. However due its toxicities in a Phase Ia/II clinical trial, AMD 3100 has been re-directed to other uses. On the other hand, AMD070, another bicyclam derivative appears to be well-tolerated and can be given orally. Also, KRH-1636 is duodenally absorbed and a potent anti-HIV-1 reagent. Recently, a low-molecular anti-HIV-1 agent with activity against both CCR5 and CXCR4, namely AMD3451 has been identified [82]. While the specific sites at which this molecule binds these HIV-1 receptors need to be clarified, it may prove to be effective both in prevention of transmission and pathogenesis of HIV-1. Whether AMD3451 will find application in management of dual HIV-1/TB remains to be investigated.

As noted, evidence for replication of both CCR5 and CXCR4 tropic HIV-1 at sites of active MTB infection during TB is accumulating. This appears to be based on the cytokine/chemokine *milieu* associated with human TB in situ. Whether the now well-documented progression of HIV-1 disease in subjects with dual HIV-1/TB is related to expansion of viral phenotype to include both CCR5 and CXCR4 remains to be clarified.

References

[1] Moore, J. P., Kitchen, S. G., Pugach, P. and Zack, J. A. The CCR5 and CXCR4 coreceptors--central to understanding the transmission and pathogenesis of human immunodeficiency virus type 1 infection *AIDS Res Hum Retroviruses* 2004 20, 111-126.

[2] Regoes, R. R. and Bonhoeffer, S. The HIV coreceptor switch: a population dynamical perspective *Trends Microbiol* 2005 13, 269-277.

[3] Briz, V., Poveda, E. and Soriano, V. HIV entry inhibitors: mechanisms of action and resistance pathways *J Antimicrob Chemother* 2006 57, 619-627.

[4] Doms, R. W. Chemokine receptors and HIV entry *Aids* 2001 15 Suppl 1, S34-35.

[5] Connor, R. I., Sheridan, K. E., Ceradini, D., Choe, S. and Landau, N. R. Change in coreceptor use coreceptor use correlates with disease progression in HIV-1--infected individuals *J Exp Med* 1997 185, 621-628.

[6] Zhang, Y. J. and Moore, J. P. Will multiple coreceptors need to be targeted by inhibitors of human immunodeficiency virus type 1 entry? *J Virol* 1999 73, 3443-3448.

[7] Paxton, W. A., Dragic, T., Koup, R. A. and Moore, J. P. The beta-chemokines, HIV type 1 second receptors, and exposed uninfected persons *AIDS Res Hum Retroviruses* 1996 12, 1203-1207.

[8] Galvani, A. P. and Novembre, J. The evolutionary history of the CCR5-Delta32 HIV-resistance mutation *Microbes Infect* 2005 7, 302-309.

[9] Sheppard, H. W., Celum, C., Michael, N. L., O'Brien, S., Dean, M., Carrington, M., Dondero, D. and Buchbinder, S. P. HIV-1 infection in individuals with the CCR5-Delta32/Delta32 genotype: acquisition of syncytium-inducing virus at seroconversion *J Acquir Immune Defic Syndr* 2002 29, 307-313.

[10] Michael, N. L., Nelson, J. A., KewalRamani, V. N., Chang, G., O'Brien, S. J., Mascola, J. R., Volsky, B., Louder, M., White, G. C., 2nd, Littman, D. R., Swanstrom, R. and O'Brien, T. R. Exclusive and persistent use of the entry coreceptor CXCR4 by human

immunodeficiency virus type 1 from a subject homozygous for CCR5 delta32 *J Virol* 1998 72, 6040-6047.

[11] Dean, M., Carrington, M., Winkler, C., Huttley, G. A., Smith, M. W., Allikmets, R., Goedert, J. J., Buchbinder, S. P., Vittinghoff, E., Gomperts, E., Donfield, S., Vlahov, D., Kaslow, R., Saah, A., Rinaldo, C., Detels, R. and O'Brien, S. J. Genetic restriction of HIV-1 infection and progression to AIDS by a deletion allele of the CKR5 structural gene. Hemophilia Growth and Development Study, Multicenter AIDS Cohort Study, Multicenter Hemophilia Cohort Study, San Francisco City Cohort, ALIVE Study [see comments] [published erratum appears in Science 1996 Nov 15;274(5290):1069] *Science* 1996 273, 1856-1862.

[12] Lee, B., Sharron, M., Montaner, L. J., Weissman, D. and Doms, R. W. Quantification of CD4, CCR5, and CXCR4 levels on lymphocyte subsets, dendritic cells, and differentially conditioned monocyte-derived macrophages *Proc Natl Acad Sci U S A* 1999 96, 5215-5220.

[13] Harouse, J. M., Gettie, A., Tan, R. C., Blanchard, J. and Cheng-Mayer, C. Distinct pathogenic sequela in rhesus macaques infected with CCR5 or CXCR4 utilizing SHIVs *Science* 1999 284, 816-819.

[14] Agace, W. W., Amara, A., Roberts, A. I., Pablos, J. L., Thelen, S., Uguccioni, M., Li, X. Y., Marsal, J., Arenzana-Seisdedos, F., Delaunay, T., Ebert, E. C., Moser, B. and Parker, C. M. Constitutive expression of stromal derived factor-1 by mucosal epithelia and its role in HIV transmission and propagation *Curr Biol* 2000 10, 325-328.

[15] Rich, E. A., Chen, I. S., Zack, J. A., Leonard, M. L. and O'Brien, W. A. Increased susceptibility of differentiated mononuclear phagocytes to productive infection with human immunodeficiency virus-1 (HIV-1) *J Clin Invest* 1992 89, 176-183.

[16] Shimizu, N., Haraguchi, Y., Takeuchi, Y., Soda, Y., Kanbe, K. and Hoshino, H. Changes in and discrepancies between cell tropisms and coreceptor uses of human immunodeficiency virus type 1 induced by single point mutations at the V3 tip of the env protein *Virology* 1999 259, 324-333.

[17] Verrier, F., Borman, A. M., Brand, D. and Girard, M. Role of the HIV type 1 glycoprotein 120 V3 loop in determining coreceptor usage *AIDS Res Hum Retroviruses* 1999 15, 731-743.

[18] Richman, D. D. and Bozzette, S. A. The impact of the syncytium-inducing phenotype of human immunodeficiency virus on disease progression *J Infect Dis* 1994 169, 968-974.

[19] Hu, Q. X., Barry, A. P., Wang, Z. X., Connolly, S. M., Peiper, S. C. and Greenberg, M. L. Evolution of the human immunodeficiency virus type 1 envelope during infection reveals molecular corollaries of specificity for coreceptor utilization and AIDS pathogenesis *J Virol* 2000 74, 11858-11872.

[20] Kimata, J. T., Gosink, J. J., KewalRamani, V. N., Rudensey, L. M., Littman, D. R. and Overbaugh, J. Coreceptor specificity of temporal variants of simian immunodeficiency virus Mne *J Virol* 1999 73, 1655-1660.

[21] Kwa, D., Vingerhoed, J., Boeser, B. and Schuitemaker, H. Increased in vitro cytopathicity of CC chemokine receptor 5-restricted human immunodeficiency virus type 1 primary isolates correlates with a progressive clinical course of infection *J Infect Dis* 2003 187, 1397-1403.

[22] Shankarappa, R., Margolick, J. B., Gange, S. J., Rodrigo, A. G., Upchurch, D., Farzadegan, H., Gupta, P., Rinaldo, C. R., Learn, G. H., He, X., Huang, X. L. and Mullins, J. I. Consistent viral evolutionary changes associated with the progression of human immunodeficiency virus type 1 infection *J Virol* 1999 73, 10489-10502.

[23] Singh, A. and Collman, R. G. Heterogeneous spectrum of coreceptor usage among variants within a dualtropic human immunodeficiency virus type 1 primary-isolate quasispecies *J Virol* 2000 74, 10229-10235.

[24] Cho, M. W., Lee, M. K., Carney, M. C., Berson, J. F., Doms, R. W. and Martin, M. A. Identification of determinants on a dualtropic human immunodeficiency virus type 1 envelope glycoprotein that confer usage of CXCR4 *J Virol* 1998 72, 2509-2515.

[25] Tersmette, M., Gruters, R. A., de Wolf, F., de Goede, R. E., Lange, J. M., Schellekens, P. T., Goudsmit, J., Huisman, H. G. and Miedema, F. Evidence for a role of virulent human immunodeficiency virus (HIV) variants in the pathogenesis of acquired immunodeficiency syndrome: studies on sequential HIV isolates *J Virol* 1989 63, 2118-2125.

[26] Troyer, R. M., Collins, K. R., Abraha, A., Fraundorf, E., Moore, D. M., Krizan, R. W., Toossi, Z., Colebunders, R. L., Jensen, M. A., Mullins, J. I., Vanham, G. and Arts, E. J. Changes in human immunodeficiency virus type 1 fitness and genetic diversity during disease progression *J Virol* 2005 79, 9006-9018.

[27] Aung, H., McKenna, S. M., Ketoff, N. R., Jones, L., Wu, M., Hejal, R., Rich, E. A. and Toossi, Z. Dysregulation of beta-chemokines in the lungs of HIV-1-infected patients *J Acquir Immune Defic Syndr* 2001 26, 305-314.

[28] Guillon, J. M., Autran, B., Denis, M., Fouret, P., Plata, F., Mayaud, C. M. and Akoun, G. M. Human immunodeficiency virus-related lymphocytic alveolitis *Chest* 1988 94, 1264-1270.

[29] Harries, A. D., Hargreaves, N. J., Chimzizi, R. and Salaniponi, F. M. Highly active antiretroviral therapy and tuberculosis control in Africa: synergies and potential *Bull World Health Organ* 2002 80, 464-469.

[30] Whalen, C., Horsburgh, C. R., Hom, D., Lahart, C., Simberkoff, M. and Ellner, J. Accelerated course of human immunodeficiency virus infection after tuberculosis *Am J Respir Crit Care Med* 1995 151, 129-135.

[31] Nunn, P., Brindle, R., Carpenter, L., Odhiambo, J., Wasunna, K., Newnham, R., Githui, W., Gathua, S., Omwega, M. and McAdam, K. Cohort study of human immunodeficiency virus infection in patients with tuberculosis in Nairobi, Kenya. Analysis of early (6-month) mortality *Am Rev Respir Dis* 1992 146, 849-854.

[32] Lucas, S. and Nelson, A. M., *in* "Tuberculosis: Pathogenesis, Protection, and Control," *ASM Press*, 1994.

[33] Whalen, C. C., Nsubuga, P., Okwera, A., Johnson, J. L., Hom, D. L., Michael, N. L., Mugerwa, R. D. and Ellner, J. J. Impact of pulmonary tuberculosis on survival of HIV-infected adults: a prospective epidemiologic study in Uganda *Aids* 2000 14, 1219-1228.

[34] Toossi, Z., Mayanja-Kizza, H., Hirsch, C. S., Edmonds, K. L., Spahlinger, T., Hom, D. L., Aung, H., Mugyenyi, P., Ellner, J. J. and Whalen, C. W. Impact of tuberculosis (TB) on HIV-1 activity in dually infected patients *Clin Exp Immunol* 2001 123, 233-238.

[35] Collins, K. R., Mayanja-Kizza, H., Sullivan, B. A., Quinones-Mateu, M. E., Toossi, Z. and Arts, E. J. Greater diversity of HIV-1 quasispecies in HIV-infected individuals with active tuberculosis *J Acquir Immune Defic Syndr* 2000 24, 408-417.

[36] Collins, K. R., Quinones-Mateu, M. E., Wu, M., Luzze, H., Johnson, J. L., Hirsch, C., Toossi, Z. and Arts, E. J. Human immunodeficiency virus type 1 (HIV-1) quasispecies at the sites of Mycobacterium tuberculosis infection contribute to systemic HIV-1 heterogeneity *J Virol* 2002 76, 1697-1706.

[37] Toossi, Z. Virological and immunological impact of tuberculosis on human immunodeficiency virus type 1 disease *J Infect Dis* 2003 188, 1146-1155.

[38] Kizza, H. M., Rodriguez, B., Quinones-Mateu, M., Mirza, M., Aung, H., Yen-Lieberman, B., Starkey, C., Horter, L., Peters, P., Baseke, J., Johnson, J. L. and Toossi, Z. Persistent replication of human immunodeficiency virus type 1 despite treatment of pulmonary tuberculosis in dually infected subjects *Clin Diagn Lab Immunol* 2005 12, 1298-1304.

[39] Sadek, M. I., Sada, E., Toossi, Z., Schwander, S. K. and Rich, E. A. Chemokines induced by infection of mononuclear phagocytes with mycobacteria and present in lung alveoli during active pulmonary tuberculosis *Am J Respir Cell Mol Biol* 1998 19, 513-521.

[40] Fantuzzi, L., Conti, L., Gauzzi, M. C., Eid, P., Del Corno, M., Varano, B., Canini, I., Belardelli, F. and Gessani, S. Regulation of chemokine/cytokine network during in vitro differentiation and HIV-1 infection of human monocytes: possible importance in the pathogenesis of AIDS *J Leukoc Biol* 2000 68, 391-399.

[41] Mayanja-Kizza, H., Wajja, A., Wu, M., Peters, P., Nalugwa, G., Mubiru, F., Aung, H., Vanham, G., Hirsch, C., Whalen, C., Ellner, J. and Toossi, Z. Activation of beta-Chemokines and CCR5 in Persons Infected with Human Immunodeficiency Virus Type 1 and Tuberculosis *J Infect Dis* 2001 183, 1801-1804.

[42] Toossi, Z., Mayanja-Kizza, H., Baseke, J., Peters, P., Wu, M., Abraha, A., Aung, H., Okwera, A., Hirsch, C. and Arts, E. Inhibition of human immunodeficiency virus-1 (HIV-1) by beta-chemokine analogues in mononuclear cells from HIV-1-infected patients with active tuberculosis *Clin Exp Immunol* 2005 142, 327-332.

[43] Eckstein, D. A., Penn, M. L., Korin, Y. D., Scripture-Adams, D. D., Zack, J. A., Kreisberg, J. F., Roederer, M., Sherman, M. P., Chin, P. S. and Goldsmith, M. A. HIV-1 actively replicates in naive CD4(+) T cells residing within human lymphoid tissues *Immunity* 2001 15, 671-682.

[44] Hoshino, Y., Tse, D. B., Rochford, G., Prabhakar, S., Hoshino, S., Chitkara, N., Kuwabara, K., Ching, E., Raju, B., Gold, J. A., Borkowsky, W., Rom, W. N., Pine, R. and Weiden, M. Mycobacterium tuberculosis-induced CXCR4 and chemokine expression leads to preferential X4 HIV-1 replication in human macrophages *J Immunol* 2004 172, 6251-6258.

[45] Toossi, Z., Johnson, J. L., Kanost, R. A., Wu, M., Luzze, H., Peters, P., Okwera, A., Joloba, M., Mugyenyi, P., Mugerwa, R. D., Aung, H., Ellner, J. J. and Hirsch, C. S. Increased Replication of HIV-1 at Sites of Mycobacterium tuberculosis Infection: Potential Mechanisms of Viral Activation *J Acquir Immune Defic Syndr* 2001 28, 1-8.

[46] Nakata, K., Rom, W. N., Honda, Y., Condos, R., Kanegasaki, S., Cao, Y. and Weiden, M. *Mycobacterium tuberculosis* enhances human immunodeficiency virus-1 replication in the lung *Am J Respir Crit Care Med* 1997 155, 996-1003.

[47] Richter, C., Perenboom, R., Mtoni, I., Kitinya, J., Chande, H., Swai, A. B., Kazema, R. R. and Chuwa, L. M. Clinical features of HIV-seropositive and HIV-seronegative

patients with tuberculous pleural effusion in Dar es Salaam, Tanzania *Chest* 1994 106, 1471-1475.

[48] Zhang, Y., Nakata, K., Weiden, M. and Rom, W. N. Mycobacterium tuberculosis enhances human immunodeficiency virus-1 replication by transcriptional activation at the long terminal repeat *J Clin Invest* 1995 95, 2324-2331.

[49] Schutze, S., Machleidt, T. and Kronke, M. Mechanisms of tumor necrosis factor action *Semin Oncol* 1992 19, 16-24.

[50] Fauci, A. S. Host factors and the pathogenesis of HIV-induced disease *Nature* 1996 384, 529-534.

[51] Toossi, Z. Cytokine circuits in tuberculosis *Infect Agents Dis* 1996 5, 98-107.

[52] Hirsch, C. S., Toossi, Z., Johnson, J. L., Luzze, H., Ntambi, L., Peters, P., McHugh, M., Okwera, A., Joloba, M., Mugyenyi, P., Mugerwa, R. D., Terebuh, P. and Ellner, J. J. Augmentation of Apoptosis and Interferon-gamma Production at Sites of Active Mycobacterium tuberculosis Infection in Human Tuberculosis *J Infect Dis* 2001 183, 779-788.

[53] Losana, G., Bovolenta, C., Rigamonti, L., Borghi, I., Altare, F., Jouanguy, E., Forni, G., Casanova, J. L., Sherry, B., Mengozzi, M., Trinchieri, G., Poli, G., Gerosa, F. and Novelli, F. IFN-gamma and IL-12 differentially regulate CC-chemokine secretion and CCR5 expression in human T lymphocytes *J Leukoc Biol* 2002 72, 735-742.

[54] Lederman, M. M., Georges, D. L., Kusner, D. J., Mudido, P., Giam, C. Z. and Toossi, Z. Mycobacterium tuberculosis and its purified protein derivative activate expression of the human immunodeficiency virus *J Acquir Immune Defic Syndr* 1994 7, 727-733.

[55] Toossi, Z., Nicolacakis, K., Xia, L., Ferrari, N. A. and Rich, E. A. Activation of latent HIV-1 by Mycobacterium tuberculosis and its purified protein derivative in alveolar macrophages from HIV-infected individuals in vitro *J Acquir Immune Defic Syndr Hum Retrovirol* 1997 15, 325-331.

[56] Garrait, V., Cadranel, J., Esvant, H., Herry, I., Morinet, P., Mayaud, C. and Israel-Biet, D. Tuberculosis generates a microenvironment enhancing the productive infection of local lymphocytes by HIV *J Immunol* 1997 159, 2824-2830.

[57] Zhang, M., Gong, J., Iyer, D. V., Jones, B. E., Modlin, R. L. and Barnes, P. F. T cell cytokine responses in persons with tuberculosis and human immunodeficiency virus infection *J Clin Invest* 1994 94, 2435-2442.

[58] Mayanja-Kizza, H., Jones-Lopez, E., Okwera, A., Wallis, R. S., Ellner, J. J., Mugerwa, R. D. and Whalen, C. C. Immunoadjuvant prednisolone therapy for HIV-associated tuberculosis: a phase 2 clinical trial in Uganda *J Infect Dis* 2005 191, 856-865.

[59] Wallis, R. S., Nsubuga, P., Whalen, C., Mugerwa, R. D., Okwera, A., Oette, D., Jackson, J. B., Johnson, J. L. and Ellner, J. J. Pentoxifylline therapy in human immunodeficiency virus-seropositive persons with tuberculosis: a randomized, controlled trial *J Infect Dis* 1996 174, 727-733.

[60] Naruse, K., Ueno, M., Satoh, T., Nomiyama, H., Tei, H., Takeda, M., Ledbetter, D. H., Coillie, E. V., Opdenakker, G., Gunge, N., Sakaki, Y., Iio, M. and Miura, R. A YAC contig of the human CC chemokine genes clustered on chromosome 17q11.2 *Genomics* 1996 34, 236-240.

[61] Scott, H. M. and Flynn, J. L. Mycobacterium tuberculosis in chemokine receptor 2-deficient mice: influence of dose on disease progression *Infect Immun* 2002 70, 5946-5954.

[62] Schwander, S. K., Sada, E., Torres, M., Escobedo, D., Sierra, J. G., Alt, S. and Rich, E. A. T lymphocytic and immature macrophage alveolitis in active pulmonary tuberculosis *J Infect Dis* 1996 173, 1267-1272.

[63] Toossi, Z., Sierra-Madero, J. G., Blinkhorn, R. A., Mettler, M. A. and Rich, E. A. Enhanced susceptibility of blood monocytes from patients with pulmonary tuberculosis to productive infection with human immunodeficiency virus type 1 *J Exp Med* 1993 177, 1511-1516.

[64] Orenstein, J. M., Fox, C. and Wahl, S. M. Macrophages as a source of HIV during opportunistic infections *Science* 1997 276, 1857-1861.

[65] Weiden, M., Tanaka, N., Qiao, Y., Zhao, B. Y., Honda, Y., Nakata, K., Canova, A., Levy, D. E., Rom, W. N. and Pine, R. Differentiation of monocytes to macrophages switches the Mycobacterium tuberculosis effect on HIV-1 replication from stimulation to inhibition: modulation of interferon response and CCAAT/enhancer binding protein beta expression *J Immunol* 2000 165, 2028-2039.

[66] Lawn, S. D., Pisell, T. L., Hirsch, C. S., Wu, M., Butera, S. T. and Toossi, Z. Anatomically compartmentalized human immunodeficiency virus replication in HLA-DR+ cells and CD14+ macrophages at the site of pleural tuberculosis coinfection *J Infect Dis* 2001 184, 1127-1133.

[67] Toossi, Z., Gogate, P., Shiratsuchi, H., Young, T. and Ellner, J. J. Enhanced production of TGF-beta by blood monocytes from patients with active tuberculosis and presence of TGF-beta in tuberculous granulomatous lung lesions *J Immunol* 1995 154, 465-473.

[68] Langermans, J. A., Nibbering, P. H., Van Vuren-Van Der Hulst, M. E. and Van Furth, R. Transforming growth factor-beta suppresses interferon-gamma-induced toxoplasmastatic activity in murine macrophages by inhibition of tumour necrosis factor-alpha production *Parasite Immunol* 2001 23, 169-175.

[69] Princen, K. and Schols, D. HIV chemokine receptor inhibitors as novel anti-HIV drugs *Cytokine Growth Factor Rev* 2005 16, 659-677.

[70] Jiang, S., Lin, K., Strick, N. and Neurath, A. R. Inhibition of HIV-1 infection by a fusion domain binding peptide from the HIV-1 envelope glycoprotein GP41 *Biochem Biophys Res Commun* 1993 195, 533-538.

[71] Robertson, D. US FDA approves new class of HIV therapeutics *Nat Biotechnol* 2003 21, 470-471.

[72] Kilby, J. M. and Eron, J. J. Novel therapies based on mechanisms of HIV-1 cell entry *N Engl J Med* 2003 348, 2228-2238.

[73] Rimsky, L. T., Shugars, D. C. and Matthews, T. J. Determinants of human immunodeficiency virus type 1 resistance to gp41-derived inhibitory peptides *J Virol* 1998 72, 986-993.

[74] Hartley, O., Gaertner, H., Wilken, J., Thompson, D., Fish, R., Ramos, A., Pastore, C., Dufour, B., Cerini, F., Melotti, A., Heveker, N., Picard, L., Alizon, M., Mosier, D., Kent, S. and Offord, R. Medicinal chemistry applied to a synthetic protein: development of highly potent HIV entry inhibitors *Proc Natl Acad Sci U S A* 2004 101, 16460-16465.

[75] Lederman, M. M., Veazey, R. S., Offord, R., Mosier, D. E., Dufour, J., Mefford, M., Piatak, M., Jr., Lifson, J. D., Salkowitz, J. R., Rodriguez, B., Blauvelt, A. and Hartley, O. Prevention of vaginal SHIV transmission in rhesus macaques through inhibition of CCR5 *Science* 2004 306, 485-487.

[76] Pastore, C., Picchio, G. R., Galimi, F., Fish, R., Hartley, O., Offord, R. E. and Mosier, D. E. Two mechanisms for human immunodeficiency virus type 1 inhibition by N-terminal modifications of RANTES *Antimicrob Agents Chemother* 2003 47, 509-517.

[77] Torre, V. S., Marozsan, A. J., Albright, J. L., Collins, K. R., Hartley, O., Offord, R. E., Quinones-Mateu, M. E. and Arts, E. J. Variable sensitivity of CCR5-tropic human immunodeficiency virus type 1 isolates to inhibition by RANTES analogs *J Virol* 2000 74, 4868-4876.

[78] Kawamura, T., Bruse, S. E., Abraha, A., Sugaya, M., Hartley, O., Offord, R. E., Arts, E. J., Zimmerman, P. A. and Blauvelt, A. PSC-RANTES blocks R5 human immunodeficiency virus infection of Langerhans cells isolated from individuals with a variety of CCR5 diplotypes *J Virol* 2004 78, 7602-7609.

[79] Mosier, D. E., Picchio, G. R., Gulizia, R. J., Sabbe, R., Poignard, P., Picard, L., Offord, R. E., Thompson, D. A. and Wilken, J. Highly potent RANTES analogues either prevent CCR5-using human immunodeficiency virus type 1 infection in vivo or rapidly select for CXCR4-using variants *J Virol* 1999 73, 3544-3550.

[80] Marozsan, A. J., Torre, V. S., Johnson, M., Ball, S. C., Cross, J. V., Templeton, D. J., Quinones-Mateu, M. E., Offord, R. E. and Arts, E. J. Mechanisms involved in stimulation of human immunodeficiency virus type 1 replication by aminooxypentane RANTES *J Virol* 2001 75, 8624-8638.

[81] Lim, J. K., Lu, W., Hartley, O. and Devico, A. L. N-terminal proteolytic processing by cathepsin G converts RANTES/CCL5 and related analogs into a truncated 4-68 variant *J Leukoc Biol* 2006.

[82] Princen, K., Hatse, S., Vermeire, K., Aquaro, S., De Clercq, E., Gerlach, L. O., Rosenkilde, M., Schwartz, T. W., Skerlj, R., Bridger, G. and Schols, D. Inhibition of human immunodeficiency virus replication by a dual CCR5/CXCR4 antagonist *J Virol* 2004 78, 12996-13006.

In: Chemokine Research Trends
Editor: Leah R. Grinwald, pp. 61-76

ISBN: 978-60021-588-9

Chapter 4

ROLE OF CHEMOKINES IN RHEUMATIC DISORDERS

Hiroshi Okamoto**, *Shigeki Momohara and Naoyuki Kamatani
Institute of Rheumatology, Tokyo Women's Medical University, Tokyo, Japan

ABSTRACT

Rheumatoid arthritis (RA) is a chronic, symmetric polyarticular joint disease characterized by synovial proliferation and the progressive destruction of cartilage and bone. We found that the expression of pulmonary and activation-regulated chemokine (PARC)/CCL18 and monocyte chemoattractant protein-4 (MCP-4)/CCL13 were significantly higher in cartilage from RA patients than from osteoarthritis (OA) patients or normal controls. Although the biological function of CCL18/PARC has not been elucidated, we found high expression in chondrocytes and synovial cells of RA patients, and determined that serum CCL18/PARC levels significantly correlated with rheumatoid factor. With regard to MCP-4/CCL13, we found that IFN-γ significantly stimulated MCP-4/CCL13 production from human chondrocytes in a dose-dependent manner and that this production was significantly enhanced in combination with IL-1β or TNF-α. MCP-4/CCL13 induces the phosphorylation of ERK in fibroblast-like synoviocytes (FLS) and activates the proliferation of these cells. A specific inhibitor of ERK kinase, PD98059, completely inhibited the stimulatory effect of MCP-4/CCL13 on FLS proliferation. From these data we concluded that IFN-γ in combination with IL-1β or TNF-α activates the production of MCP-4/CCL13 from chondrocytes in RA joints, and secreted MCP-4/CCL13 enhances FLS proliferation by activating the ERK MAP kinase cascade resulting in RA pathogenesis.

Systemic lupus erythematosus (SLE) is an autoimmune disease characterized by widespread immunologic abnormalities and multiorgan involvement, including the skin, joints, kidney, serosal membranes, and peripheral and central nervous system (CNS). Neuropsychiatric systemic lupus erythematosus (NPSLE) is a major contributor to morbidity and mortality in patients with (SLE). Our results showed that cerebral spinal fluid (CSF) levels of Th-1 type chemokines such as IP-10/CXCL10 and MCP-1/CCL2 were higher in NPSLE patients than in non-NPSLE patients. This suggests that CSF IP-10/CXCL10 and MCP-1/CCL2 reflect inflammatory activity in the brain, and may be used as a diagnostic tool as well as a gauge for monitoring therapeutic responses in patients with NPSLE.

---

* E-mail: hokamoto@ior.twmu.ac.jp Tel. +81-3-5269-1711; Fax +81-3-5269-1726 Institute of Rheumatology, Tokyo Women's Medical University, 10-22 Kawada-cho, Shinjuku-ku, Tokyo, 162-0054, Japan

INTRODUCTION

Rheumatoid arthritis (RA) and systemic lupus erythematosus (SLE) are representative diseases of rheumatic disorders. RA is a chronic, symmetric polyarticular joint disease characterized by synovial proliferation and progressive destruction of cartilage and bone. SLE is an autoimmune disease characterized by widespread immunologic abnormalities and multiorgan involvement, including the skin, joints, kidney, serosal membranes, and peripheral and central nervous system (CNS). In this chapter we will examine the roles of chemokines in the pathogenesis of rheumatic disorders such as RA and SLE.

Chemokines in Rheumatoid Arthritis

Rheumatoid arthritis (RA) is a chronic inflammatory disease characterized by heavy infiltration of leukocytes into the synovial tissue. Chemokines and chemokine receptors have been reported to play an important role in cell migration and positioning of leukocytes within inflamed rheumatoid synovium. Currently, there is much focus on the specific contribution and role of chemokines and chemokine receptors in the chronic inflammatory process in synovial tissue [1, 2]. Chemokines are chemotactic cytokines that regulate leukocyte migration during inflammatory responses, as well as homeostatic trafficking of lymphocytes and dendritic cells [3, 4]. Recently, the importance of the role of chemokines in autoimmune diseases, neoplasia, and cardiovascular diseases has been demonstrated [4]. Chemokines are members of a complex superfamily composed of at least 50 low-molecular weight proteins (6-14 kDa). Their primary structure is characterized by the presence of four conserved cysteine residues. Based on the positioning of the two NH2-terminal cysteines, the chemokine family can be structurally divided into CC chemokines, CXC chemokines, CX3C chemokines and C chemokines. Chemokines activate leukocytes via G-protein coupled receptors, which mediate most of their various biological activities [5]. The CC chemokines, such as monocyte chemoattractant protein-1 (MCP-1)/CCL2, macrophage inflammatory protein-1α (MIP-1α)/CCL3, MIP-1β/CCL4, and RANTES/CCL5, are implicated in RA pathogenesis via recruitment and retention of monocytes and T lymphocytes into the joints [6]. In addition to cell traffic regulation, several chemokines have been reported to enhance the migration and proliferation of fibroblast-like synoviocytes (FLS) and to up-regulate gelatinase and collagenase production by FLS [7]. Thus, chemokines are key molecules in the pathogenesis of RA and are potential therapeutic targets for RA.

We performed a DNA oligonucleotide microarray analysis to search for differentially expressed genes in RA. Significantly increased levels of CCL18/PARC and MCP-4/CCL13 mRNA, as well as highly abundant protein levels, were found in cartilage and synovial tissue from RA patients compared to that from osteoarthritis (OA) patients. Therefore, we assessed the role of CCL18/PARC and MCP-4/CCL13 in the pathology of RA.

Role of CCL18/PARC in Rheumatoid Arthritis

CCL18/PARC is a newly described CC chemokine [8] that is structurally related to CCL3/MIP-1 α, a well-characterized proinflammatory chemokine that attracts many leukocytic cell types through binding to at least two receptors, i.e. CCR1 and CCR5 [4, 9-11]. However, the CCL18/PARC receptors and the mouse equivalent for CCL18/PARC have not yet been identified. CCL18/PARC is either constitutively expressed or induced in monocytes/macrophages and dendritic cells, and chemoattracts T and B lymphocytes [12, 13]. CCL18/PARC has high structural similarity (64% identical amino acids) to the well characterized CC chemokine MIP-1α, and shares its capacity to suppress proliferation of subsets of immature myeloid progenitor cells [14]. Thus far, CCL18/PARC C expression has been reported in unrelated pathologies such as atherosclerosis [15], hypersensitivity pneumonitis [16], allergic contact hypersensitivity [17], and ovarian carcinoma [18].

There have been only two reports concerning CCL18/PARC and arthritis [18.19]. Schutyser et al. reported four-fold enhancement of CCL18/PARC levels in synovial fluids from RA patients compared to those with crystal-induced arthritis and osteoarthritis. In addition, immunochemistry revealed that CD68+ monocytes/macrophages were the main CCL18/PARC-producing cell type in arthritic synovial tissue [8]. Radstake et al. assessed the

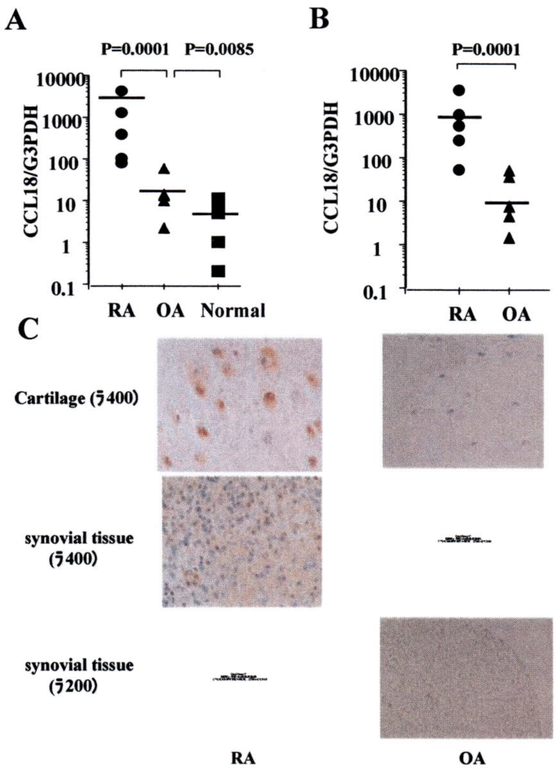

Figure 1. **(A)** Comparative expression of genes for CCL18/PARC in RA, OA and normal cartilage. CCL18/PARC expression was higher in RA cartilage than from OA or normal cartilage (P = 0.001). The expression level of each gene was normalized to the level of G3PDH mRNA. Means are represented by the horizontal bars. **(B)** Comparative expression of CCL18/PARC mRNA in RA and OA synovial tissue. CCL18/PARC expression was higher in RA synovial tissue than from OA synovial tissue (P = 0.001). The expression level of each gene was normalized to the level of G3PDH expression. Means are represented by the horizontal bars. **(C)** Immunohistologic analysis of the distribution of CCL18/PARC in cartilage and synovial tissue from patients with RA and OA.

expression of various chemokines by immature and mature dendritic cells and examined their regulation by the Fc gamma receptor in RA patients and healthy donors. They provided evidence for elevated CCL18 expression by dendritic cells in RA patients and demonstrated

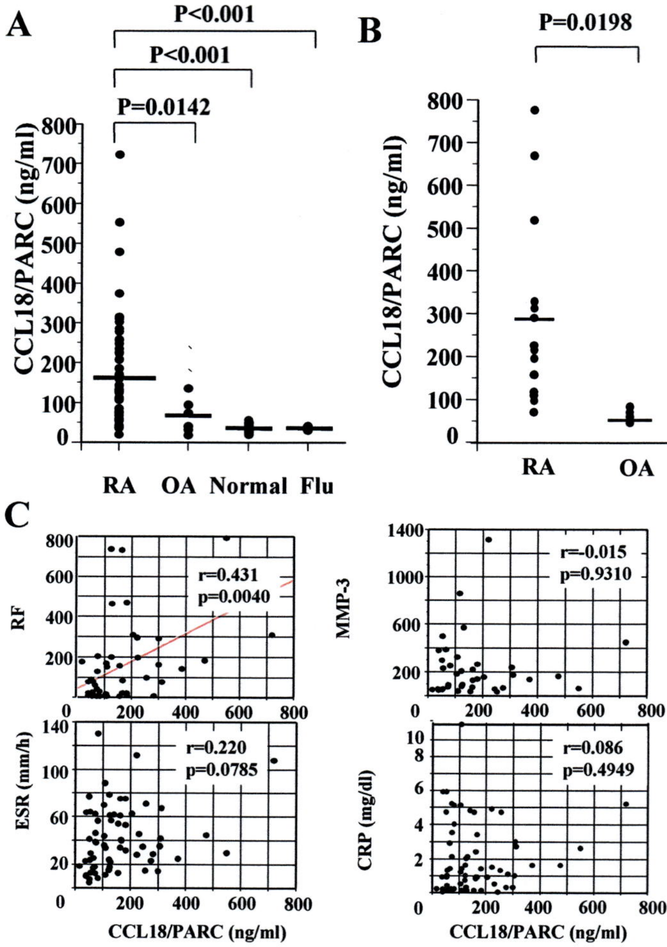

Figure 2. (**A**) CCL18/PARC serum levels in patients with RA. CCL18/PARC levels in the serum were measured by ELISA from 71 RA, 12 OA, 20 normal, and 8 influenza patients, as well as an age matched control group of patients from the same hospital. Horizontal bars indicate the median CCL18/PARC concentration for each group. Statistical analysis was performed using the Mann-Whitney U test. Significance levels for differences in CCL18/PARC serum concentration between the patient group and the control and influenza group are indicated at the top of the panels. (**B**) CCL18/PARC levels in synovial fluid of patients with RA. CCL18/PARC levels in synovial fluid were measured by ELISA from 15 RA and 6 OA patients. The horizontal bars indicate the median CCL18/PARC concentration for each group. Statistical analysis was performed using the Mann-Whitney U test. Significance levels for differences in CCL18/PARC synovial fluid concentration between RA patients and OA patients are observed (P=0.0198). (**C**) Correlation between ESR, CRP, MMP-3 and serum CCL18/PARC. CCL18/PARC levels correlated with RF (r=0.431, p=0.0040). However, assessment of the relationship between ESR, CRP, MMP-3 and CCL18/PARC levels indicated no significant correlation (CCL18 vs. ESR r=0.220, p=0.0785; CCL18 vs. CRP r=0.086, p=0.4949; CCL18 vs. MMP-3 r=-0.015, p=0.9310).

that this expression was regulated in part by FcγR triggering. They also suggested a potential role for chemokines, as well as CCL18/PARC producing dendritic cells, in RA pathogenesis [19]. In this study, CCL18/PARC mRNA was expressed at significantly higher levels in cartilage from RA patients compared to OA and control patients ($P < 0.001$). Furthermore, cartilage tissue from patients with OA showed significantly higher expression of CCL18/PARC than control patients ($P = 0.0085$). CCL18/PARC mRNA was also detected at much higher levels in synovial membranes from RA patients than from OA patients ($P = 0.0001$) (Figure 1A, B).

Immunohistochemical study revealed that all cartilage and synovial tissue samples from RA patients exhibited medium to strong staining for CCL18/PARC (Figure 1C). In OA synovial tissue there was weak staining for CCL18/PARC protein within the subintimal regions, but CCL18/PARC was undetectable in all OA and normal cartilage (Figure 1C).

Local over-expression of CCL18/PARC in the joints, cartilage and synovial tissues, may be detected by the presence of CCL18/PARC in the synovial fluid and serum in patients with RA. To this end, we analyzed CCL18/PARC in the synovial fluid and serum from patients with RA as well as OA. There were significantly high levels of CCL18/PARC (156.21 ± 125.73 ng/ml, n=71) in the serum of RA patients compared to OA patients (64.54 ± 40.90 ng/ml, n=12) and healthy donors (28.04 ± 10.96 ng/ml, n=20) (Figure 2A). However, in influenza patients with systemic inflammation and increased serum C-reactive protein (CRP), serum levels of CCL18/PARC were comparable to the levels in healthy donors (27.54 ± 1.54 ng/ml, n=8). Moreover, when CCL18/PARC was measured in the synovial fluid of RA and OA patients, the average CCL18/PARC concentrations (275.20 ± 228.16 ng/ml, n = 15) were significantly higher in the RA patients than the OA patients (33.13 ± 14.84 ng/ml, n = 6) (Figure 2B).

Role of MCP-4/CCL13 in Rheumatoid Arthritis

MCP-4/CCL13 is a CC chemokine recently identified from a human cDNA library that directs the migration of eosinophils, monocytes, and T lymphocytes through several chemokine receptors, including CCR-2 and CCR-3 [21, 22]. The role of MCP-4/CCL13 in disease is not well defined, but recent studies have suggested that it is involved in inflammatory cell recruitment in allergic disorders such as asthma and atopic dermatitis [23-26]. We analyzed the concentrations of MCP-4/CCL13 protein in synovial fluid obtained from RA patients (n=44) and OA patients (n=30) by enzyme linked immunosorbent assay (ELISA). Of the 44 RA patients, 86.4% were female and 84.1% were rheumatoid factor positive. The serum CRP concentrations of these patients were 1.32 ± 1.79 mg/dl. The concentration of MCP-4/CCL13 protein in synovial fluid from RA patients (mean 208.07 ± 462.6 pg/ml) was significantly higher than from OA patients (mean 29.6 ± 45.7 pg/ml) (p=0.0000112) (Figure 3A). Immunohistochemical study revealed that MCP-4/CCL13 protein was expressed in cartilage from RA patients, but was not detected in cartilage from OA patients (Figure 3B).

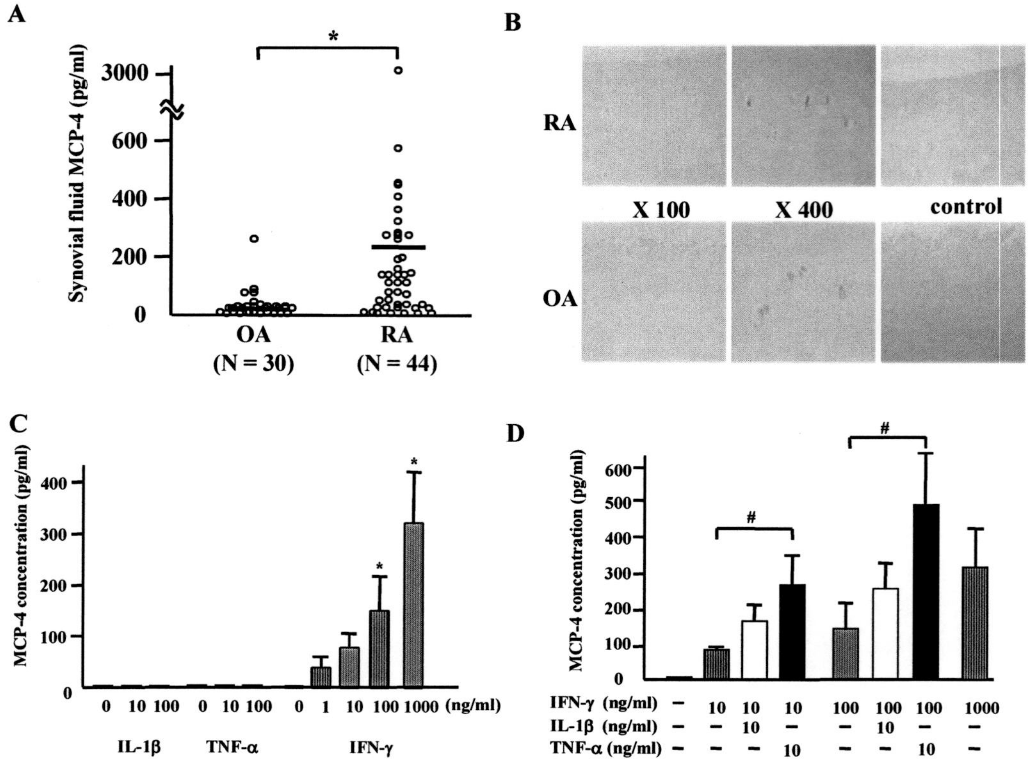

Figure 3(**A**). Synovial fluid levels of MCP-4/CCL13 in patients with RA or OA. Means are indicated by solid lines. Statistical comparisons were performed using the nonparametric Mann-Whitney U-test. * = $P < 0.01$. (**B**) Immunohistochemical staining with anti-human MCP-4/CCL13 goat polyclonal antibody. Significant expression of MCP-4/CCL13 protein was detected in cartilage from RA patients, but not in cartilage from OA patients. Results shown are representative of three RA patients and three OA patients. MCP-4/CCL13 protein production by human chondrocytes from 3 RA patients and 1 OA patient. (**C**) Human chondrocytes were cultured in the presence of IL-1β (0-100 ng/ml; open bars), TNF-α (0-100 ng/ml; solid bars) and IFN-γ (0-1000 ng/ml; shaded bars) for 48 hours. (**D**) Human chondrocytes were cultured in the presence of IFN-γ (shaded bars), IFN-γ + IL-1 β (open bars), and IFN-γ + TNF-α (solid bars) for 48 hours. MCP-4/CCL13 protein concentrations in these cell culture supernatants were evaluated by ELISA. The bars show the mean and SD for four separate experiments. Statistical evaluation was performed using one-way ANOVA followed by Tukey's method for multiple comparisons. * = $P < 0.01$ compared with vehicle-treated control. # = $P < 0.01$ compared with the sample cultured with IFN-γ alone.

To identify the stimulatory signals that activate production of MCP-4/CCL13 from human chondrocytes, we investigated the effect of several cytokines reported to play a role in RA pathogenesis. Human chondrocytes from RA or OA patients were cultured in the presence of IL-1β, TNF-α or IFN-γ, and various combinations of these three cytokines. MCP-4/CCL13 protein concentrations in culture supernatants were evaluated by ELISA. IFN-γ significantly stimulated MCP-4/CCL13 production in a dose-dependent manner, while IL-1β and TNF-α had no significant effect (Figure 3C). Interestingly, MCP-4/CCL13 production by

IFN-γ was significantly and remarkably enhanced when combined with IL-1β or TNF-α (Figure 3D).

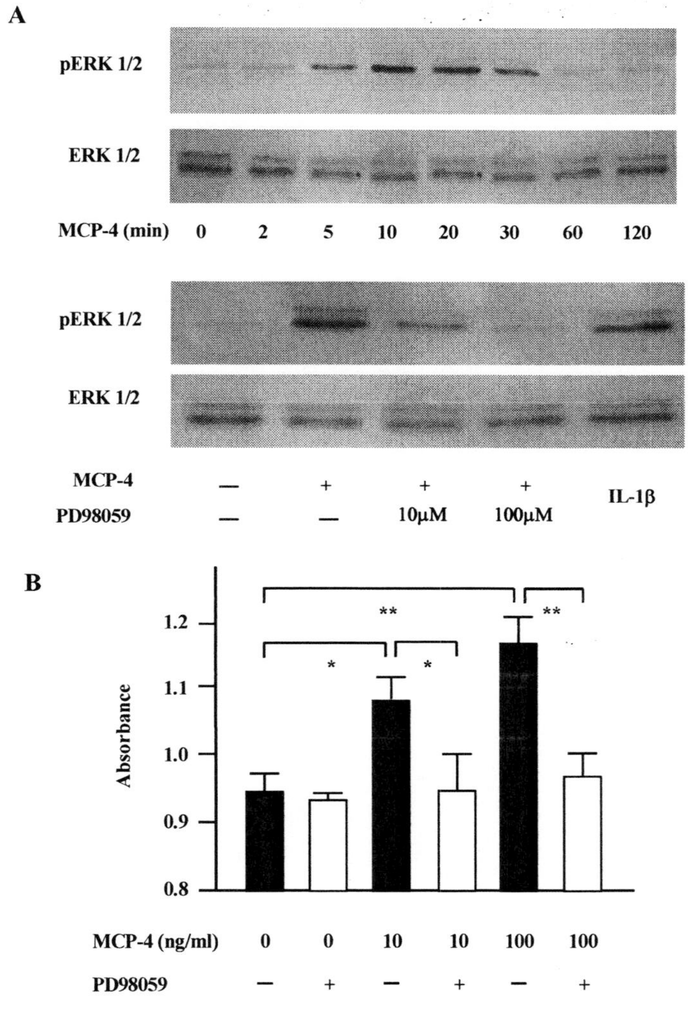

Figure 4. Induction of ERK phosphorylation by MCP-4/CCL13 in RA FLS. FLS were cultured overnight in serum-free DMEM. (**A**) FLS were incubated in the presence of MCP-4/CCL13 (100 ng/ml) for an additional 0-120 minutes. FLS were incubated in the presence of MCP-4/CCL13 (100 ng/ml) for 20 minutes with or without PD98059 (10-100 μg/ml). As a positive control, FLS were incubated with IL-1β (5 ng/ml) for 20 minutes. Cell lysates were examined for ERK activation by Western blotting with phospho-p44/42 MAPK mouse monoclonal antibody (pERK1/2). Total p44/42 MAP kinase antibody was used to verify equal protein loading. The result is one representative example from three independent experiments. (**B**) Effects of inhibition of ERK phosphorylation on RA FLS proliferation. FLS were treated with MCP-4/CCL13 (0-100 ng/ml) with and without PD98059 (100 μg/ml) for 48 hours. MCP-4/CCL13 significantly increased RA FLS proliferation, and inhibition of ERK phosphorylation by PD98059 significantly inhibited FLS proliferation. PD98059 did not inhibit basal proliferation of RA FLS. The bars show the mean and SD for three independent experiments. Statistical evaluation was performed using one-way ANOVA followed by Tukey's method for multiple comparisons.* = $P < 0.05$, ** = $P < 0.01$.

Western blot analysis was performed to investigate whether MCP-4/CCL13 could activate extracellular signal-regulated kinase (ERK), which is widely known to play a major role in cell proliferation (27). As expected, phosphorylation of ERK was induced by stimulation with MCP-4/CCL13 in RA FLS. This activation peaked around 10-20 minutes and returned to basal levels within 60 minutes. Incubation with 100 μM PD98059, the specific ERK activation inhibitor, was sufficient to abolish ERK activation by MCP-4/CCL13 in RA FLS (Figure 4A).

To confirm the involvement of ERK phosphorylation in the proliferative effect of MCP-4/CCL13 on RA FLS, an 3-(4,5-demethylthiazol-2-yl)-2,5-diphenyltetrazolium bromide (MTT) cell proliferation assay was performed using PD98059 as an antagonist. Since representative results of Western blot analyses were obtained at a concentration of 100 ng/ml of PD98059, the MTT cell proliferation assay was performed using the same concentration.

MCP-4/CCL13 enhanced the proliferation of FLS in a dose-dependent manner, and PD98059 completely inhibited the stimulatory effect of MCP-4/CCL13 on FLS proliferation. PD98059 did not inhibit basal proliferation of RA FLS (Figure 4B). These results demonstrated that MCP-4/CCL13 secreted from chondrocytes in the joints plays an important role in the development of aggressive synovial tissues in RA. Our data shows that not only synovial cells but also chondrocytes are actively involved in RA pathogenesis, and indicate that MCP-4/CCL13 may serve as a new target for anti-RA therapy [28, 29].

Role of CCR5 in Rheumatoid Arthritis

The chemokine receptor CCR5 is preferentially expressed on Th1 lymphocytes. There have been reports that RA is a T helper type 1 (Th1) cell dominant disorder, and that CCR5 and/or CXCR3 expressing cells are enriched in affected joints of RA [30, 31]. Yang et al. demonstrated that the CCR5 antagonist inhibits the development of arthritis in a collagen induced arthritis (CIA) mouse model [32]. They showed that inhibition of arthritis development is not caused by an effect on generation of collagen-sensitized T cells, but by interfering with migration to joint lesions. Recently, Vierboom et al. reported that the CCR5 antagonist inhibited development of CIA in rhesus monkeys (33). They showed that systemic administration of a small molecular weight antagonist of CCR5, SCH-X, suppressed the development of CIA in the monkey model of RA. We studied the effects of systemic administration of TAK-779, a nonpeptide compound with a small molecular weight (*M*r 531.13) [34], on the development of adjuvant-induced arthritis in rats. Arthritis was stimulated by injection of 100 μL of Complete Freund's adjuvant (CFA) emulsion intradermally at the base of the left paw. TAK779 was obtained from Takeda Co., Ltd. (Osaka, Japan). Intravenous administration of TAK-779 or PBS (control) at the specified dose was given once a day, commencing at the onset of arthritis and continuing for 7 days. As shown in Figure 5, TAK-779 suppressed the progression of clinical arthritis compared with control rats treated with PBS, as demonstrated by paw volume and arthritis score (Figure 5A). In the TAK-779-treated rats, statistically significant effects were observed at higher doses ($P<0.01$). By day 18, histological analysis of the ankle joint in TAK-779 rats treated at 1 mg/kg/day and 2 mg/kg/day, and control rats given PBS, showed that TAK-779 inhibited mononuclear cell infiltration and pannus formation in synovial tissue (Figure 5B). Histology of a normal rat joint at the same age is shown in Figure 1B. There was no mortality and no body weight loss in the TAK-779-treated rats. These data suggest that TAK-779 has anti-arthritis effects *in vivo* and further supports the evidence shown by others that CCR5 plays an important role in the development of arthritis in animal models of human RA. Thus, therapy using a CCR5 antagonist may serve as a new strategy for treatment of RA [35].

A

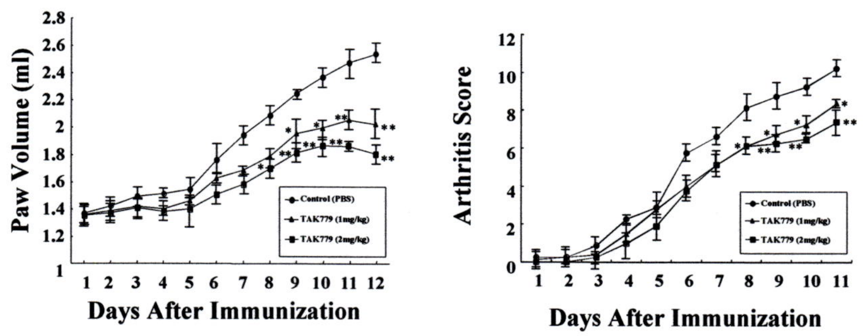

B

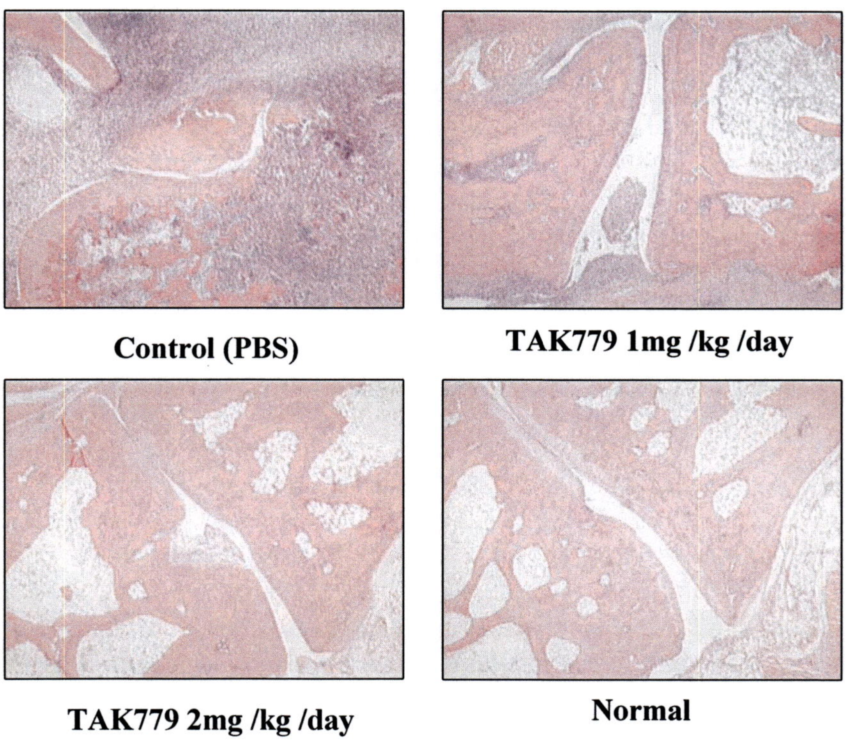

Figure 5. Suppression of arthritis development by TAK-779 in an adjuvant-induced arthritis model. (**A**) TAK-779 suppressed the progression of clinical arthritis compared to control rats treated with PBS, as demonstrated by paw volume and arthritis score. The data are represented as mean ± standard deviation. * P< 0.05, ** P< 0.01. (**B**) Histological findings of the foot joint in normal rats, TAK-779 -treated rats (1 mg/kg/day), TAK-779 -treated rats (2 mg /kg /day) and control rats (PBS).

CHEMOKINES IN SYSTEMIC LUPUS ERYTHEMATOSUS

SLE has long been considered to be a disease in which the actions of peripheral T helper type 2 cells (Th2) predominate over Th1 cells [36, 37]. However, this notion was recently challenged by a study that reported a predominance of Th1 cells in SLE patients with World Health Organization (WHO) class IV lupus nephritis [38]. Therefore, the relative balance of Th1 versus Th2 cells in SLE patients remains somewhat controversial.

There have been reports showing that serum levels of MCP-1/CCL2 are higher in SLE patients than healthy controls [39, 40]. Previously, we reported that there were significant differences in plasma concentrations of TARC/CCL17 between patients with untreated SLE, treated SLE ($p<0.001$), RA ($p<0.001$), and healthy controls ($p<0.001$) [41]. In addition, plasma levels of TARC/CCL17 are dependent upon the class of lupus nephritis, i.e. levels are higher in class I or II than class III or IV disease. CCR4+ CD4+ T cells (Th2 cells) have been reported to accumulate in the kidneys of patients with lupus nephritis, but were less prevalent in the peripheral blood of these patients, suggesting that chemokines actively recruit CCR4+ cells to inflamed renal tissue [42]. Our results are consistent with this observation since patients with lupus nephritis exhibited elevated concentrations of circulating TARC/CCL17, which may facilitate recruitment of CCR4+ CD4+ T cells to renal tissues. Therefore, we speculated that MCP-1/CCL2 plays an important role in cellular immunity at a later stage in the pathogenesis of SLE than the stage in which TARC/CCL17 contributes.

Neuropsychiatric systemic lupus erythematosus (NPSLE) may occur at any time during the course of the disease, and symptoms are extremely diverse, ranging from depression to psychosis, seizures, or stroke [43]. The origin of minor clinical symptoms, such as headaches and mood swings, are difficult to differentiate, and SLE patients may be affected by other conditions capable of causing neuropsychiatric symptoms such as infection, severe hypertension, metabolic derangement, steroid psychosis or other drug toxicities [44]. Without proper treatment, neuropsychiatric involvement in SLE may increase morbidity and mortality. Therefore, the availability of beneficial treatment also increases the need for early recognition of neuropsychiatric manifestations. In addition to more specific diagnostic tools, an effective method of monitoring disease activity and therapeutic responses is crucial in the management of NPSLE.

Currently, diagnostic procedures for NPSLE include brain magnetic resonance imaging (MRI), electroencephalogram (EEG), neuropsychological testing, and lumbar puncture. These findings are reported to be abnormal in some but not all patients with NPSLE, and therefore, are not specific indicators of NSPLE. The large discrepancy in the reported frequency of neuropsychiatric involvement in SLE patients (14%-75%) provides further indication that there is no single confirmatory diagnostic tool [45, 46]. Proinflammatory cytokines have been known to be increased in the cerebral spinal fluid (CSF) of patients with NPSLE, and some reports have shown that cytokines such as interleukin-6 (IL-6), IL-1, IL-8, IL-10, tumor necrosis factor (TNF)–alpha, and interferon (IFN)-gamma are elevated intrathecally, allowing these cytokines to be used as diagnostic tools [47, 48]. However, reports on the role of chemokines in NPSLE are limited. MCP-1/CCL2 is mainly expressed by activated monocyte/macrophages, T cells, and natural killer cells, and it attracts leukocytes and other mediators to sites of inflammation. Studies showing the involvement of NPSLE and MCP-1/CCL2 are also scarce. It is a high-affinity ligand for the CCR2 chemokine receptor that is mainly expressed on Th1 cells. The predominance of Th1 versus Th2 cells in NPSLE patients remains unresolved.

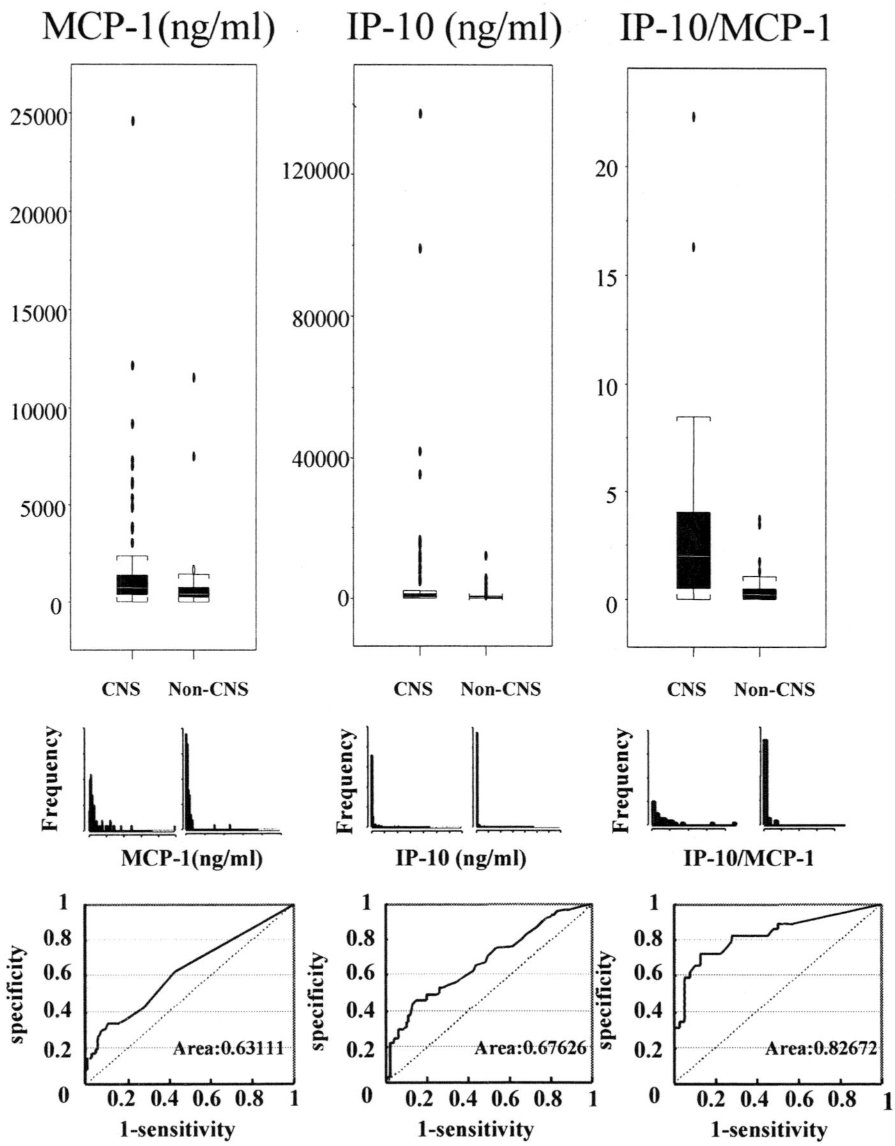

Figure 6(A). CSF levels of IP-10/CXCL10 in SLE patients with (NPSLE) and without (non-NPSLE) neuropsychiatric involvement. (**B**) CSF levels of MCP-1/CCL2 in SLE patients with (NPSLE) and without (non-NPSLE) neuropsychiatric involvement. (**C**) CSF IP-10/MCP-1 ratios in SLE patients with (NPSLE) and without (non-NPSLE) neuropsychiatric involvement. The discriminative ability (area under the curve) of each group was 0.63111 (95% confidence interval [95% CI] 822.4064-7787.1975) (IP-10/CXCL10); 0.67626 ([95% CI] 418.6142-1761.8262) (MCP-1/CCL2); and 0.82672 ([95% CI] 1.222729-5.011448) (IP-10/MCP-1 ratio).

Interferon inducible protein-10 (IP-10)/CXCL10 and (MCP-1)/CCL2 are high-affinity ligands for the chemokine receptors CXCR3 and CCR2 respectively. To find a useful diagnostic marker of NPSLE, we measured the concentration of IP-10/CXCL10 and MCP-

1/CCL2 in CSF from NPSLE and non-NPSLE patients. A total of 202 NPSLE patients fulfilling the criteria defined by the American College of Rheumatology were chosen for this study. Patients were divided into two groups: 101 patients were SLE patients with CNS symptoms (NPSLE), and 101 patients were SLE patients without CNS symptoms (non-NPSLE). Cerebrospinal fluid was collected from these patients and IP-10/CXCL10 and MCP-1/CCL2 concentrations were evaluated by enzyme linked immunoassay (ELISA). The NPSLE patients had active NPSLE manifestations when CSF samples were obtained and were treated with systemic corticosteroid alone. No samples were obtained after extensive treatments such as corticosteroid pulse and/or cyclophosphamide pulse therapy. The average concentration of IP-10/CXCL10 was 3748.78 $\pm$ 15543.68 pg/ml in NPSLE patients and 483.82 $\pm$ 1433.67 pg/ml in non-NPSLE patients (Figure 6A). The average concentration of MCP-1/CCL2 was 1757.68 $\pm$ 3135.09 pg/ml in NPSLE patients and 667.46 $\pm$ 1349.05 pg/ml in non-NPSLE patients (Figure 6B). Statistical analyses using the Mann-Whitney U test revealed that the IP-10/CXCL10 and MCP-1/CCL2 concentrations in the NPSLE group were significantly higher than those in the non-NPSLE group (P=0.0000137 and P=0.0000150, respectively). Interestingly, the IP-10/MCP-1 ratio in the NPSLE group was significantly higher than that in the non-NPSLE group (P=0.0000014) (Figure 6C). The discriminative ability (area under the curve) of each chemokine was 0.63111 (95% confidence interval [95% CI] 822.4064-7787.1975) (IP-10/CXCL10); 0.67626([95% CI] 418.6142-1761.8262) (MCP-1/CCL2); and 0.82672 ([95% CI] 1.222729-5.011448) (IP-10/MCP-1 ratio). There were no correlations between the levels of MCP-1 in CSF and serum (r=-1332, p=0.2481) and the levels of IP-10 in CSF and serum (r=-1445, p=0.3842). There were no correlations between the levels of MCP-1 and IL-6 (r=-0.1229, p=0.462), IL-8 (r=-0.1585, p=0.349) and IFN alpha (r=-0.0818, p=0.179); and between IP-10 and IL-6 (r=-0.1435, p=0.384), IL-8 (r=-0.1553, p=0.332) and IFN alpha (r=-0.1021, p=0.213). CXCR3, the receptor of IP-10/CXCL10, is predominantly expressed on activated T cells and natural killer cells. On the other hand, CCR2, the receptor of MCP-1/CCL2, is expressed not only on activated T cells and natural killer cells, but also on monocytes, basophils, and dendritic cells. These results indicate a differential contribution of CXCR3 and CCR2 signaling in the pathogenesis of NPSLE. We concluded that Th1 chemokines have an important role in the pathogenesis of NPSLE, and showed that CSF IP-10/MCP-1 ratios are higher in NPSLE patients than in non-NPSLE patients, indicating that this index is a useful diagnostic marker of NPSLE [49-51].

CONCLUSION

In this study, experimental and clinical data provided insight into the role of chemokines in the pathogenesis of RA and SLE. Our results demonstrated that CCL18/PARC and CCR5 have important roles in the pathogenesis of RA, presumably through activation of Th1 cells. Our data also suggested that MCP-4/CCL13 from RA chondrocytes play an important role in synovial proliferation, which is a typical pathological feature of RA. In cases of SLE, Th2 chemokines such as TARC/CCL17 play a role in the early stages of SLE, and Th1 chemokines play an important role in the later stages. This is evident in the CNS systems of NPSLE patients where Th1 chemokines such as MCP-1/CCL2 and IP-10CXCL10 have prerequisite roles. New therapeutic strategies using synthetic compounds or biologics that act as chemokine antagonists may serve as a new strategy in the treatment of RA and SLE.

Although these strategies are promising, further experimental and clinical data must be accumulated on the potential risks of biologics prior to clinical use [52, 53].

REFERENCES

[1] Gay, S; Gay, RE; Koopman, WJ. Molecular and cellular mechanisms of joint destruction in rheumatoid arthritis: two cellular mechanisms explain joint destruction? *Ann Rheum Dis*, 1993 52, S39-47.

[2] Muller-Ladner, U. Molecular and cellular interactions in rheumatoid synovium. *Curr Opin Rheumatol*, 1996 8, 210-20.

[3] Loetscher, P; Moser, B; Baggiolini, M. Chemokines and their receptors in lymphocyte traffic and HIV infection. *Adv Immunol*, 2000 74, 127-80.

[4] Gerard, C; Rollins, BJ. Chemokines and disease. *Nat Immunol*, 2001 2, 108-15.

[5] Murphy, PM; Baggiolini, M; Charo, IF; Hébert, CA; Horuk, R; Matsushima, K; Miller, LH; Oppenheim, JJ; Power, CA. International union of pharmacology. XXII. Nomenclature for chemokine receptors. *Pharmacol Rev*, 2000 52, 145-76.

[6] Patel, DD; Zachariah, JP; Whichard, LP. CXCR3 and CCR5 ligands in rheumatoid arthritis synovium. *Clin Immunol*, 2001 98, 39-45.

[7] García-Vicuña, R; Gómez-Gaviro, MV; Domínguez-Luis, MJ; Pec, MK; Gonzalez-Alvaro, I; Alvaro-Gracia, JM; Diaz-Gonzalez, F. CC and CXC chemokine receptors mediate migration, proliferation, and matrix metalloproteinase production by fibroblast-like synoviocytes from rheumatoid arthritis patients. *Arthritis Rheum*, 2004 50, 3866-77.

[8] Schutyser, E; Struyf, S; Wuyts, A; Put, W; Geboes, K; Grillet, B; Opdenakker, G; Van Damme, J. Selective induction of CCL18/PARC by staphylococcal enterotoxins in mononuclear cells and enhanced levels in septic and rheumatoid arthritis. *Eur J Immunol*, 2001 31, 3755-62.

[9] Hieshima, K; Imai, T; Baba, M; Shoudai, K; Ishizuka, K; Nakagawa, T; Tsuruta, J; Takeya, M; Sakaki, Y; Takatsuki, K; Miura, R; Opdenakker, G; Van Damme, J; Yoshie, O; Nomiyama, H. A novel human CC chemokine PARC that is most homologous to macrophage-inflammatory protein-1 alpha/LD78 alpha and chemotactic for T lymphocytes, but not for monocytes. *J. Immunol*, 1997 159, 1140-9.

[10] Kodelja, V; Müller, C; Politz, O; Hakij, N; Orfanos, CE; Goerdt, S. Alternative macrophage activation-associated CC-chemokine-1, a novel structural homologue of macrophage inflammatory protein-1 alpha with a Th2-associated expression pattern. *J. Immunol*, 1998 160, 1411-8.

[11] Guan, P; Burghes, AH; Cunningham, A; Lira, P; Brissette, WH; Neote, K; McColl, SR. Genomic organization and biological characterization of the novel human CC chemokine DC-CK-1/PARC/MIP-4/SCYA18. *Genomics*, 1999 56, 296-302.

[12] Adema, GJ; Hartgers, F; Verstraten, R; de Vries, E; Marland, G; Menon, S; Foster, J; Xu, Y; Nooyen, P; McClanahan, T; Bacon, KB; Figdor, CG. A dendritic-cell-derived C-C chemokine that preferentially attracts naive T cells. *Nature*, 1997 387, 713-7.

[13] Lindhout, E; Vissers, JL; Hartgers, FC; Huijbens, RJ; Scharenborg, NM; Figdor, CG; Adema, GJ. The dendritic cell-specific CC-chemokine DC-CK1 is expressed by

germinal center dendritic cells and attracts CD38-negative mantle zone B lymphocytes. *J Immunol*, 2001 166, 3284-9.

[14] Broxmeyer, HE; Kim, CH; Cooper, SH; Hangoc, G; Hromas, R; Pelus, LM. Effects of CC, CXC, C, and CX3C chemokines on proliferation of myeloid progenitor cells, and insights into SDF-1-induced chemotaxis of progenitors. *Ann N Y Acad Sci*, 1999 872, 142-62.

[15] Reape, TJ; Rayner, K; Manning, CD; Gee, AN; Barnette, MS; Burnand, KG; Groot, PH. Expression and cellular localization of the CC chemokines PARC and ELC in human atherosclerotic plaques. *Am J Pathol*, 1999 154, 365-74.

[16] Pardo, A; Smith, KM; Abrams, J; Coffman, R; Bustos, M; McClanahan, TK; Grein, J; Murphy, EE; Zlotnik, A; Selman, M. CCL18/DC-CK-1/PARC up-regulation in hypersensitivity pneumonitis. *J Leukoc Biol*, 2001 70, 610-6.

[17] Goebeler, M; Trautmann, A; Voss, A; Bröcker, EV; Toksoy, A; Gillitzer, R. Differential and sequential expression of multiple chemokines during elicitation of allergic contact hypersensitivity. *Am J Pathol*, 2001 158, 431-40.

[18] Schutyser, E; Struyf, S; Proost, P; Opdenakker, G; Laureys, G; Verhasselt, B; Peperstraete, L; Van de Putte, I; Saccani, A; Allavena, P; Mantovani, A; Van Damme, J. Identification of biologically active chemokine isoforms from ascitic fluid and elevated levels of CCL18/pulmonary and activation-regulated chemokine in ovarian carcinoma. *J Biol Chem*, 2002 277, 24584-93.

[19] Radstake, TR; van der Voort, R; ten Brummelhuis, M; de Waal Malefijt, M; Looman, M; Figdor, CG; van den Berg, WB; Barrera, P; Adema, GJ. Increased expression of CCL18, CCL19, and CCL17 by dendritic cells from patients with rheumatoid arthritis and regulation by Fc gamma receptors. *Ann Rheum Dis*, 2005 64, 359-67.

[20] Momohara, S; Okamoto, H; Iwamoto, T; Mizumura, T; Ikari, K; Kawaguchi, Y; Takeuchi, M; Kamatani, N; Tomatsu, T. High CCL18/PARC Expression in Articular Cartilage and Synovial Tissue of Rheumatoid Arthritis. *J Rheumatol, in press*.

[21] Garcia-Zepeda, EA: Combadiere, C; Rothenberg, ME; Sarafi, MN; Lavigne, F; Hamid, Q; Murphy, PM; Luster, AD. Human monocyte chemoattractant protein (MCP)-4 is a novel CC chemokine with activities on monocytes, eosinophils, and basophils induced in allergic and nonallergic inflammation that signals through the CC chemokine receptors (CCR)-2 and -3. *J Immunol*, 1996 157, 5613-26.

[22] Uguccioni, M; Loetscher, P; Forssmann, U; Dewald, B; Li, H; Lima, SH; Li, Y; Kreider, B; Garotta, G; Thelen, M; Baggiolini, M. Monocyte chemotactic protein 4 (MCP-4), a novel structural and functional analogue of MCP-3 and eotaxin. *J Exp Med*, 1996 183, 2379-84.

[23] Lamkhioued, B; Garcia-Zepeda, EA; Abi-Younes, S; Nakamura, H; Jedrzkiewicz, S; Wagner, L; Renzi, PM; Allakhverdi, Z, Lilly, C; Hamid, Q; Luster, AD. Monocyte chemoattractant protein (MCP)-4 expression in the airways of patients with asthma. Induction in epithelial cells and mononuclear cells by proinflammatory cytokines. *Am J Respir Crit Care Med*, 2000 162, 723-32.

[24] Kalayci, O; Sonna, LA; Woodruff, PG; Camargo, CA Jr.; Luster, AD; Lilly, CM. Monocyte chemotactic protein-4 (MCP-4; CCL-13): a biomarker of asthma. *J Asthma*, 2004 41, 27-33.

[25] Taha, RA; Minshall, EM; Miotto, D; Shimbara, A; Luster, A; Hogg, JC; Hamid, QA. Eotaxin and monocyte chemotactic protein-4 mRNA expression in small airways of asthmatic and nonasthmatic individuals. *J Allergy Clin Immunol*, 1999 103, 476-83.

[26] Taha, RA; Minshall, EM; Leung, DY; Boguniewicz, M; Luster, A; Muro, S; Toda, M; Hamid, QA. Evidence for increased expression of eotaxin and monocyte chemotactic protein-4 in atopic dermatitis. *J Allergy Clin Immunol*, 2000 105, 1002-7.

[27] Kolch, W. Meaningful relationships: the regulation of the Ras/Raf/MEK/ERK pathway by protein interactions. *Biochem J*, 2000 351, 289-305.

[28] Iwamoto, T; Okamoto, H; Iikuni, N; Takeuchi, M; Toyama, Y; Tomatsu, T; Kamatani, N; Momohara, S. Monocyte chemoattractant protein-4 (MCP-4)/CCL13 is highly expressed in cartilage from patients with rheumatoid arthritis. *Rheumatology (Oxford)*, 2006 45, 421-4.

[29] Iwamoto, T; Okamoto, H; Ikari, K; Kobayashi, S; Toyama, Y; Tomatsu, T; Kamatani, N; Momohara, S. A role of Monocyte Chemoattractant Protein-4 (MCP-4)/CCL13 from chondrocytes in rheumatoid arthritis. In submission.

[30] Simchen, C; Lehmann, I; Sittig, D; Steinert, M; Aust G. Expression and regulation of regulated on activation, normal T cells expressed and secreted in thyroid tissue of patients with Graves' disease and thyroid autonomy and in thyroid-derived cell populations. *J Clin Endocrinol Metab*, 2000 85, 4758-64.

[31] Suzuki, N; Nakajima, A; Yoshino, S; Matsushima, K; Yagita, H; Okumura, K. Selective accumulation of CCR5+ T lymphocytes into inflamed joints of rheumatoid arthritis. *Int Immunol*, 1999 11, 553-9.

[32] Yang, YF; Mukai, T; Gao, P; Yamaguchi, N; Ono, S; Iwaki, H; Obika, S; Imanishi, T; Tsujimura, T; Hamaoka, T; Fujiwara H. A non-peptide CCR5 antagonist inhibits collagen-induced arthritis by modulating T cell migration without affecting anti-collagen T cell responses. *Eur J Immunol*, 2002 32, 2124-32.

[33] Vierboom, MP; Zavodny, PJ; Chou, CC; Tagat, JR; Pugliese-Sivo, C; Strizki, J; Steensma, RW; McCombie, SW; Celebi-Paul, L; Remarque, E; Jonker, M; Narula, SK; Hart, B. Inhibition of the development of collagen-induced arthritis in rhesus monkeys by a small molecular weight antagonist of CCR5. *Arthritis Rheum*, 2005 52, 627-36.

[34] Baba, M; Nishimura, O; Kanzaki, N; Okamoto, M; Sawada, H; Iizawa, Y; Shiraishi, M; Aramaki, Y; Okonogi, K; Ogawa, Y; Meguro, K; Fujino, M. A small-molecule, nonpeptide CCR5 antagonist with highly potent and selective anti-HIV-1 activity. *Proc Natl Acad Sci U S A*, 1999 96, 5698-703.

[35] Okamoto, H; Kamatani, N. A CCR-5 antagonist inhibits the development of adjuvant arthritis in rats. *Rheumatology* (*Oxford*), 2006 45, 230-2.

[36] Caligaris-Cappio, F; Bertero, MT; Converso, M; Stacchini, A; Vinante, F; Romagnani, S; Pizzolo, G. Circulating levels of soluble CD30, a marker of cells producing Th2-type cytokines, are increased in patients with systemic lupus erythematosus and correlate with disease activity. *Clin Exp Rheumatol*, 1995 13, 339-43.

[37] Funauchi, M; Ikoma, S; Enomoto, H; Horiuchi, A. Decreased Th1-like and increased Th2-like cells in systemic lupus erythematosus. *Scand J Rheumatol*, 1998 27, 219-24.

[38] Akahoshi, M; Nakashima, H; Tanaka, Y; Kohsaka, T; Nagano, S; Ohgami, E; Arinobu, Y; Yamaoka, K; Niiro, H; Shinozaki, M; Hirakata, H; Horiuchi, T; Otsuka, T; Niho, Y. Th1/Th2 balance of peripheral T helper cells in systemic lupus erythematosus. *Arthritis Rheum*, 1999 42, 1644-8.

[39] Kaneko, H; Ogasawara, H; Naito, T; Akimoto, H; Lee, S; Hishikawa T; Sekigawa, I; Tokano, Y; Takasaki, Y; Hirose, SI; Hashimoto, H. Circulating levels of beta-chemokines in systemic lupus erythematosus. *J Rheumatol*, 1999 26, 568-73.

[40] Matsukawa, A; Kaplan, MH; Hogaboam, CM; Lukacs, NW; Kunkel, SL. Pivotal role of signal transducer and activator of transcription (Stat)4 and Stat6 in the innate immune response during sepsis. *J Exp Med*, 2001 193, 679-88.

[41] Okamoto, H; Koizumi, K; Yamanaka, H; Saito, T; Kamatani, N. A role for TARC/CCL17, a CC chemokine, in systemic lupus erythematosus. J Rheumatol, 2003 30, 2369-73.

[42] Yamada, M; Yagita, H; Inoue, H; Takanashi, T; Matsuda, H; Munechika, E; Kanamaru, Y; Shirato, I; Tomino, Y; Matushima, K; Okumura, K; Hashimoto, H. Selective accumulation of CCR4+ T lymphocytes into renal tissue of patients with lupus nephritis. *Arthritis Rheum*, 2002 46, 735-40.

[43] ACR Ad Hoc Committee on Neuropsychiatric Lupus Nomenclature. The American College of Rheumatology nomenclature and case definitions for neuropsychiatric lupus syndromes. Arthritis Rheum, 1999 42, 599-608.

[44] Bresnihan, B. CNS lupus. *Clin* Rheum *Dis*, 1982 8, 183-95.

[45] Brey, RL; Holliday, SL; Saklad, AR; Navarrete, MG; Hermosillo-Romo, D; Stallworth, CL; Valdez, CR; Escalante, A; del Rincon, I; Gronseth, G; Rhine, CB; Padilla, P; McGlasson, D. Neuropsychiatric syndromes in lupus: prevalence using standardized definitions. *Neurology*, 2002 58, 1214-20.

[46] Ainiala, H; Loukkola, J; Peltola, J; Korpela, M; Heitaharju, A. The prevalence of neuropsychiatric syndromes in systemic lupus erythematosus. *Neurology*, 2001 57, 496-500.

[47] Trysberg, E; Carlsten, H; Tarkowski, A. Intrathecal cytokines in systemic lupus erythematosus with central nervous system involvement. *Lupus*, 2000 9, 498-503.

[48] Svenungsson, E; Andersson, M; Brundin, L; van Vollenhoven, R; Khademi, M; Tarkowski, A; Greitz, D; Dahlstrom, M; Lundberg, I; Klareskog, L; Olsson, T. Increased levels of proinflammatory cytokines and nitric oxide metabolites in neuropsychiatric lupus erythematosus. *Ann Rheum Dis*, 2001 60, 372-9.

[49] Okamoto, H; Katsumata, Y; Nishimura, K; Kamatani, N. Interferon-inducible protein 10/CXCL10 is increased in the cerebrospinal fluid of patients with central nervous system lupus. *Arthritis Rheum*, 2004 50, 3731-2.

[50] Iikuni, N; Okamoto, H; Yoshio, T; Sato, E; Kamitsuji, S; Iwamoto, T; Momohara, S; Taniguchi, A; Yamanaka, H; Minota, S; Kamatani, N. Raised monocyte chemotactic protein-1 (MCP-1)/CCL2 in cerebrospinal fluid of patients with neuropsychiatric lupus. *Ann Rheum Dis*, 2006 65, 253-6.

[51] Okamoto, H; Iikuni, N; Kamitsuji, S; Yoshio, T; Minota, S; Kamatani, N. IP-10/MCP-1 ratio in CSF is an useful diagnostic marker of neuropsychiatric lupus patients. *Rheumatology* (*Oxford*), 2006 45, 232-4.

[52] Okamoto, H; Kamatani, N. Rituximab for rheumatoid arthritis. N Engl *J Med*, 2004 351, 1909-1909.

[53] Okamoto, H. Chaperonin 10 in patients with rheumatoid arthritis. *Lancet, in press*.

In: Chemokine Research Trends
Editor: Leah R. Grinwald, pp. 77-88
ISBN: 978-60021-588-9

Chapter 5

SMOKING AND CHEMOKINE LEVELS

Sumiko Iho[1*] and Takayuki Takahashi[2]
[1]Department of Pathological Sciences, Faculty of Medical Sciences, University of Fukui, Fukui, Japan
[2]Department of Hematology and Clinical Immunology, Kobe City General Hospital, Kobe, Japan.

ABSTRACT

Leukocytosis is a well-known hematological disorder in tobacco smokers. This malady includes neutrophilia, lymphocytosis, and additionally, under a certain analysis, monocytosis and eosinophilia. The cause of the leukocytosis, however, has yet to be clarified. One of the reasons may be due to the fact that it remains difficult to account for such a pattern of leukocytosis with the *in vivo* biologic action of single hematopoiteic cytokine. We have attempted to identify the cause of leukocytosis with respect to blood chemokine levels and the effect of nicotine, a major substance in tobacco smoke, on the chemokine production by white cells *in vitro*. The blood levels of IL-8, MCP-1, and RANTES were elevated in smokers. Nicotine promoted IL-8 and MCP-1 production by neutrophils as well as RANTES production by T lymphocytes. Activation of NF-κB was involved in the production of these chemokines. However, signals upstream NF-κB differed, i.e., neutrophil IL-8 production was mediated by ROIs, but ROIs were only partially involved in the RANTES production by T cells. Based on the examination for the smokers who smoke 20 cigarettes and more a day and have 10,000/μL leukocytes or more, we propose that the levels of IL-8 and RANTES in the blood could be adequate markers of heavy smoking with leukocytosis.

INTRODUCTION

Tobacco smoking exerts many unfavorable effects on almost all tissues or organs [1]. As far as the effect on the hematopoietic system is concerned, leukocytosis has been well recognized [2-4]. Leukocytosis includes neutrophilia, lymphocytosis (T cells), and, in a certain analysis [3, 4], monocytosis. The cause of leukocytosis, however, has not yet been

* E-mail address: ihosumik@u-fukui.ac.jp

clarified. Regarding cigarette smoking, it has been shown in an animal study that an accelerated release of neutrophils from the bone marrow leads to leukocytosis [5]. Therefore, IL-8 seems to be a possible candidate cytokine responsible for leukocytosis in smokers. IL-8 is produced by leukocytic and nonleukocytic cells. Among these cells, neutrophils produce a very small quantity of IL-8, but when stimulated with IL-1, IL-15, tumor necrosis factor (TNF)-α, or lipopolysaccharide (LPS), they produce large amounts of IL-8 [6]. On the other hand, while smoking tobacco, it is likely that neutrophils circulating in the blood vessels of the respiratory tracts are exposed to high concentrations of nicotine, a kind of alkaloid and a major substance in tobacco smoke. If nicotine stimulates these neutrophils to produce IL-8, leukocytosis in smokers could largely be explained by this mechanism because of the largest cell population of neutrophils in the circulating white cells.

IL-8 has previously been reported to enhance lymphocyte trafficking or recruitment to the IL-8-injected site in experimental animals [7, 8]; however, the issue is still controversial. We hypothesized that RANTES (regulated upon activation, normal T-cell expressed and secreted), a potent chemoattractant for T cells, may be responsible for lymphocytosis and to a certain extent for mixed leukocyte recruitment. We also speculated monocytes chemoattractant protein (MCP)-1 as a chemokine causing leukocytosis in smokers.

In this study, we compared the blood level of IL-8, MCP-1, and RANTES between smokers and nonsmokers and examined the effect of nicotine on the production of these chemokines *in vitro*. The mechanisms of nicotine-induced chemokine production were also investigated in relation to the activation of NF-κB, a key transcription factor for these gene expressions. Furthermore, to clarify whether these nicotine-induced chemokines contribute to leukocytosis in smokers with respect to their biological actions, in particular, to their leukocyte-mobilizing actions *in vivo*, we measured the blood levels of chemokines in heavy smokers with leukocytosis.

Materials and Methods

Cells

Mononuclear cells and neutrophils were separated from citrated venous blood of healthy volunteers, with informed consent, by a two-step gradient centrifugation with specific gravities of 1.076 g/mL (for mononuclear cells) and 1.090 g/mL (for neutrophils). T cells were purified from the mononuclear cells as $CD19^-/CD56^-/CD14^-/CD304^-$ cells. The preparation contained >98% of neutrophils or T cells; and >99% of the cells were viable.

Measurement of Chemokines, Reactive Oxygen Intermediates (ROIs), Nitric Oxide (NO), and NF-κB Activity [9]

Chemokines in blood and cell-free culture supernatants were determined using enzyme-linked immunosorbent assay kits. Their gene expressions were examined by reverse transcriptase polymerase chain reaction (RT-PCR). Superoxide anion (O_2^-) was monitored by the reduction of extracellular cytochrome c. The mean fluorescence intensities of the intracellular dihydrorhodamine 123 (DHR) were measured to evaluate the oxidative potentials of $ONOO^-$ and $ONOO^-$ production was assessed with the inhibition test using a specific NO synthase (NOS) inhibitor, N^{ω}-nitro-L-arginine methyl ester (L-NAME). Regarding NO, nitrite concentrations of the culture supernatants were measured by the Salzman method. NF-κB activities were assayed using the TransAM NFκB p50 and p65 Kits (Active Motif, Carlsbad, CA) according to the manufacturer's instruction.

Statistical Analysis

Statistical analyses were performed using Student's *t* test for parametric data, the Mann-Whitney U test for non-parametric data, and Spearman's rank correlation coefficient for analysis of correlation. Fifty % effective concentration (EC_{50}) and 50% inhibiting concentration (IC_{50}) were calculated using curve-fitting program GraphPad Prism 2.0 (GraphPad, San Diego, CA).

RESULTS

Blood Levels of IL-8 and RANTES are Elevated in Smokers

The possibility of smoking-induced chemokine production *in vivo* was examined by comparing the levels in sera between 20 smokers and 30 non-smokers from 20 to 49 years of age (male). The numbers of cigarettes smoked by smokers were from 5 to 60 a day. There was no correlation between chemokine levels and age in the non-smokers. The IL-8 levels were significantly higher at $p<0.01$ in the smokers (1.11 ± 0.09 pg/mL, mean ± SE) as compared with those of the non-smokers (0.41 ± 0.07 pg/ml). The levels of RANTES in smoker's sera (22,390 ± 3,993 pg/mL) were also higher than those in nonsmokers' sera (14,903 ± 2,298 pg/mL), although the difference was smaller than that of IL-8 because of the use of sera (p=0.1~0.05).

Nicotine Induces *in vitro* the Production of IL-8 and RANTES, in Neutrophils and Activated T Cells, Respectively

Nicotine ([-]-1-methyl-2-[3-pyridyl] pyrrolidine) induced neutrophils from healthy non-smokers to produce IL-8 in both a dose- and time-dependent manner [9]. The minimum effect of nicotine was observed at 0.1 mM, and the concentrations needed to induce maximum IL-8 release ranged from 2.5 to 3.0 mM, with an EC_{50} of 1.89 ± 0.37 mM (mean ± standard deviation, n = 6). The maximum amount produced by nicotine was equivalent to that produced by 10 ng/ml LPS. An obvious increase in nicotine-induced IL-8 production was observed from 3 h of culture. Nicotine-stimulated neutrophils did not produce TNF-α or IL-1β, and the addition of polymyxin B (LPS inhibitor) did not alter the nicotine effect, suggesting that nicotine directly triggers the production of IL-8 in neutrophils. Freshly-isolated monocytes were unresponsive to nicotine in producing IL-8. Nicotine also upregulated RANTES production in T cells, activated with anti-CD3 monoclonal antibody or not activated, in a dose-dependent manner similar to that for IL-8 (manuscript in preparation).

However, the culture as long as 12-36 h was required for the up-regulation of RANTES production. Both IL-8 and RANTES production were abrogated by the pre-treatment of cells with actinomycin D, indicating that nicotine stimulates *de novo* synthesis of IL-8 and RANTES.

Since the effective dose of nicotine for the upregulation of IL-8 and RANTES production was much higher than that observed for the central nervous systems (200~300 nM at EC_{50}), the levels of mRNA were analyzed regarding RANTES expression by activated T cells. Figure 1 shows that the physiological concentration of nicotine is effective for the up-regulation of mRNA expression. The discrepancy of the concentrations needed for protein synthesis and mRNA expression remain unknown.

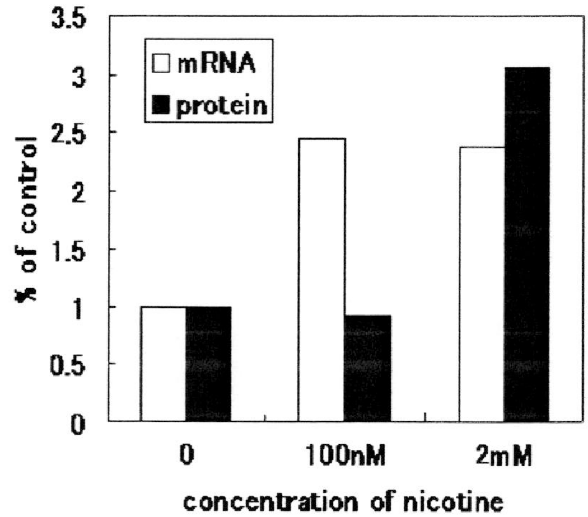

Figure 1. Nicotine up-regulates the expression of RANTES. T cells activated by anti-CD3 monoclonal antibody were cultured for 4 days with nicotine at the concentration of 100 nM or 2 mM. The mRNA for RANTES was analyzed by real-time PCR and the amount in the culture supernatant was measured by ELISA. The data are expressed as the relative values against the control.

IL-8 and RANTES Production Appears to Be Mediated by the Interaction of Nicotine with Nicotinic Acetylcholine Receptors (nAChRs)

Nicotine can bind to neutrophils [10, 11] and unprimed human T cells express nAchR (our unpublished observation). However, none of the particular nAChR subtypes responsible for nicotine-induced chemokine production have yet been determined. We tested whether the nicotine effect is mediated by nAChRs using nicotinic antagonist(s), and found that mecamylamine ((N, 2, 3, 3-tetramethyl-bicyclo [2.2.1] heptan-2-amine), a channel blocker of the $\alpha 3$- and $\alpha 4$-subtypes, inhibited the production of IL-8 and RANTES with 0.58 mM and 0.48 mM of EC_{50} against 2 mM nicotine, respectively. It is suggested that this chemokine production is mediated by nAChRs. The definite subunits involved are under investigation.

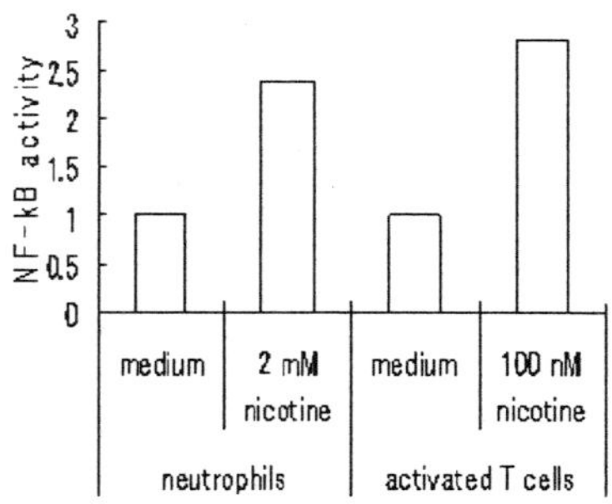

Figure 2. Nicotine up-regulates the activity of NF-κB in neutrophils and T cells. Neutrophils and CD3 Ab-activated T cells were cultured for 3 hours with 2mM and 100 nM of nicotine, respectively. Whole cell extract was analyzed for the DNA binding activity of NF-κB (p50), and the activity was expressed as relative values against the control.

Nicotine up-Regulates the DNA-Binding Activity of NF-κB

The activation of NF-κB is critical for the transcription of genes for IL-8 and RANTES. Neutrophils expressed constitutively-activated NF-κB, and nicotine increased the activity of NF-κB, which was measured by the binding of p65 and p50 to the oligonucleotide containing the NF-κB consensus site, 5'-GGGACTTTCC-3' [9]. The degree of its increase was comparable to that increased by TNF-α (40 ng/ml), which was assigned as a positive signal for the activation of NF-κB. In T cells, nicotine also enhanced the activity of NF-κB p65 and p50, in association with IkappaB kinase activation (unpublished). This enhancement was observed at the physiological concentration of 100 nM, and the degree of increase became

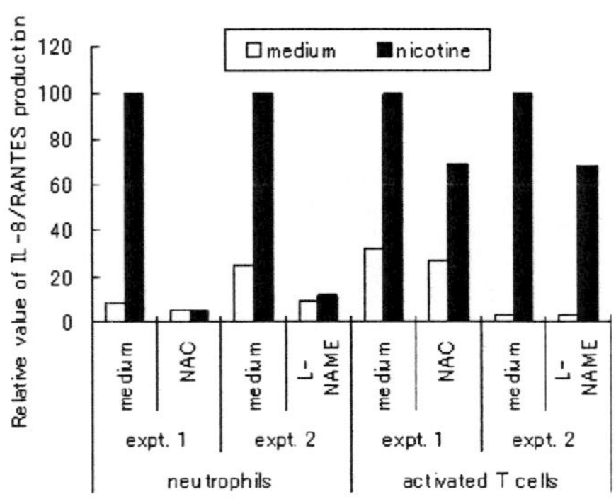

Figure 3. ROIs are involved in IL-8 production by neutrophils, but not in RANTES production by T cells. Neutrophils and T cells were cultured with medium or 2 mM nicotine in the presence or absence of NAC (20 mM) or L-NAME (10 mM), and the amount of IL-8 or RANTES in the culture supernatant was measured. The data are shown as relative values against the control culture.

large along with the culture period (e.g., 5-fold increase at 40h). Figure 2 shows the nicotine-enhanced NF-κB activity with p50 at 3h. The production of IL-8 and RANTES was completely inhibited by the pharmacological inhibitors against the NF-κB (ref. 9 for IL-8; not shown for RANTES]. The activation of NF-κB seems to be a key factor for nicotine-induced chemokine production.

Nicotine Induces $ONOO^-$ to Cause IL-8 Production in Neutrophils but Not RANTES Production in T Cells

The balance between the amounts of pro-oxidant and anti-oxidant is distorted in the blood of smokers [12], and the main source of blood free radicals in smokers is thought to be circulating neutrophils in the alveolar arterioles [13]. Since the activation of NF-κB is redox-regulated, the generation of ROI by smoking may be one of the reasons for the production of IL-8. Indeed, the IL-8 induction was abrogated in the presence of an antioxidant, N-acetyl-L-cystein (NAC) (Figure 3).

Among ROIs, $ONOO^-$ has been reported to be one of the most important initiators of IL-8 gene expression [14]. Since neutrophils are capable of generating NO both spontaneously and in response to certain stimuli, the induction of O_2^- by nicotine could possibly bring about the efficient generation of $ONOO^-$ through the reaction with NO. As expected, the NO was detected in the culture supernatant of neutrophils (<1.9 $\mu M/4x10^6$ cells/ml at 17 h); a rapid increase in the production of O_2^- was observed in nicotine-stimulated neutrophils in a 5-min culture [9]; nicotine stimulation increased intracellular DHR oxidation in neutrophils (Figure 4); and L-NAME (a specific NOS inhibitor) prevented the nicotine-enhanced DHR oxidation (Figure 4), and entirely abrogated IL-8 production (Figure 3). These results imply that $ONOO^-$ is generated by the reaction of constitutively expressed NO with nicotine-induced O_2^-

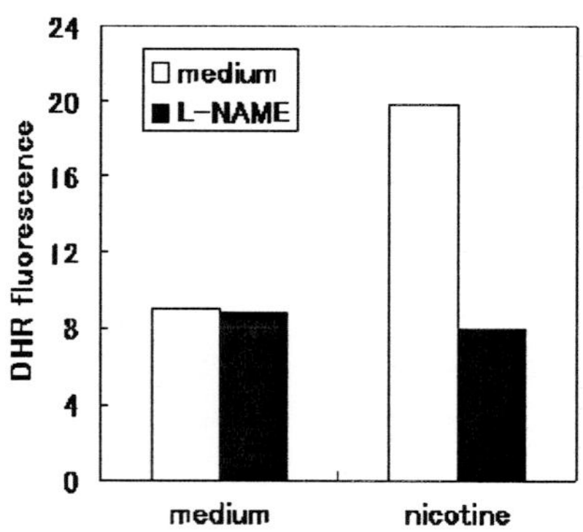

Figure 4. Nicotine up-regulates the oxidation of DHR in neutrophils. Neutrophils were cultured for 10 hours with medium or 2 mM nicotine and the mean fluorescence intensities of the intracellular DHR were measured.

and causes IL-8 production in neutrophils. It should be noted that, by contrast with the nicotine activation, the inhibition by L-NAME of TNE-α-induced IL-8 production was partial. For neutrophils, generation of ROIs, especially $ONOO^-$, appears to act as a key initiator for the nicotine-induced NF-κB activation and subsequent IL-8 production. Regarding RANTES production by T cells, however, the contribution of ROIs was only partial (Figure 3). The mechanisms activating NF-κB may be different between neutrophils and T cells stimulated with nicotine.

MCP-1 Is also Possibly Involved in Leukocytosis in Smokers

The nicotine-activation of NF-κB prompted us to consider the possibility that nicotine-stimulated neutrophils produce MCP-1 or macrophage inflammatory protein (MIP)-1α, since both require NF-κB activation for their gene transcription and recruit mixed leukocytes. The concentrations of MCP-1 in the culture supernatant of nicotine-stimulated neutrophils were 33.1 ± 10.2 pg/4 x 10^6 cells/ml (mean ± SE, $n = 5$, in 16 h-culture), which were significantly higher when compared to those in non-stimulated culture (2.4 ± 0.9). MIP-1α was also detected with 26.8 ± 12.0 in the nicotine-stimulated culture, but there was no statistical difference as compared with the amounts, 10.9 ± 4.9, in the control culture. In sera, the MCP-1 levels were higher in smokers than those in non-smokers (529 ± 23.3 pg/ml, 33 non-smokers vs. 698 ± 28.6, 21 smokers; $p < 0.01$), and MIP-1α was not detected in either group.

Blood Levels of IL-8 and RANTES Could Be Adequate Markers of Heavy Smoking

The above-mentioned data suggest that the blood levels of IL-8, RANTES, and MCP-1 could be used to evaluate the status of the heavy smoking as associated with leukocytosis. Nine smokers who smoke 20 cigarettes and more a day and have 10,000/μL leukocytes or more were evaluated for their blood-level concentration. In this group, the concentrations of IL-8 were significantly higher, ranging from 1.57 to 15.3 pg/mL (7.70 ± 1.59 pg/mL, mean ± SE), where all the values were detected above the upper level for the non-smokers (0.41 ± 0.07 pg/ml). The RANTES concentrations were measured using plasma to exclude the effect of platelets, and compared with newly obtained 15 nonsmokers' plasma where any hematological or serological aberrance was not apparent (638 ± 114, mean ± SE). In this smokers' group, the plasma RANETS ranged from 1,219 to 14,742 pg/mL (8,004 ± 1,705 pg/mL, mean ± SE), eight of nine showing levels higher than 1,567 pg/ml, which was the highest in the non-smokers' plasma tested. The concentrations of MCP-1 were not higher in such smokers' group as compared to those in non-smokers; although the elevated concentration was observed in the group consisting of smokers who take <20 cigarettes a day. Although further surveys are needed to provide a definite conclusion, the current analysis suggests that the levels of IL-8 and RANTES in blood could be adequate markers of heavy smoking with leukocytosis.

DISCUSSION

It has been shown that intravenous administration of IL-8 induces a neutrophil mobilization from the bone marrow [15, 16]. Cigarette smoking has also been suggested to contribute to neutrophilia by stimulating the bone marrow in an animal study [5]. Therefore, we presumed IL-8 to be a causative cytokine for neutrophilia in smokers. Indeed, the levels of blood IL-8, but not TNF-α, IL-1β, IL-6, granulocyte colony-stimulating factor (G-CSF), or granulocyte-macrophage CSF (data not shown), which are directly or indirectly involved in leukocytosis, were higher in smokers than in non-smokers. The positive correlation between blood IL-8 levels and heavy smoking, shown in our previous study [9], supports the *in vivo* effect of smoking on IL-8 production.

From our *in vitro* results, we submit that nicotine-activated neutrophils contribute to the elevated blood-levels of IL-8 in smokers. There have been reports showing that cigarette smoke induces IL-8 in human bronchial epithelial cells [17], and in alveolar macrophages in guinea pigs [18]. As another source of IL-8, therefore, cells in the respiratory system should be taken into consideration. Nevertheless, because neutrophils are the largest population of cells in the circulation and are concentrated in the pulmonary capillaries [19] and because monocytes did not respond to nicotine *in vitro* (our observation and [20]), IL-8 produced by nicotine in these neutrophils should greatly reflect the elevated blood-IL-8 concentration. The *in vitro* data that the nicotine-enhanced neutrophil migration was mediated by IL-8 [9] also support our hypothesis.

It has been reported that IL-8 promotes lymphocyte trafficking or recruitment to the IL-8-injected site in experimental animals [7, 8]. We observed that in the study with nude mice, subcutaneous inoculation of human carcinoma cell lines (PC-3 and HLC-1) that produce IL-8, but not IL-1, IL-6, TNF-α, or G-CSF [21], caused marked lymphocytosis, as well as neutrophilia (unpublished observations), although a single IL-8 injection appears not to induce significant lymphocytosis [22]. Furthermore, this study showed that nicotine-stimulated neutrophils produced MCP-1, and the blood MCP-1 levels were detected to be higher when analyzed in the group with various smoking habits. MCP-1 attracts monocytes and T lymphocytes [22-24]. Therefore, the neutrophil activation seems to contribute to smoker's leukocytosis, by causing not only neutrophilia but also lymphocytosis and monocytosis. Leukocytosis in smokers would be also explained by nicotine-induced RANTES production, because the blood levels of RNATES were significantly higher in heavy smokers and nicotine up-regulated RANTES production in T cells.

Nicotine is detected in smokers' sera at 25~444 nM [25]. Therefore, the gene expression may be stimulated with those concentrations of nicotine as shown in Figure 1. However, on the protein synthesis, which was evaluated by the amount secreted in the culture supernatant, the conspicuous effect was observed only at concentrations above 100 μM. It has been assumed that, in smokers, the nicotine concentrations in the tissues of the respiratory tract are as high as those in their saliva [26], in which nicotine concentrations reached mM levels [27]. Therefore, the cells pre-activated with low concentrations of nicotine could be signaled to produce large amounts of chemokines by being exposed to effective concentrations of nicotine, while passing through the lung or residing in the tissues of the respiratory tract. Another explanation for the discrepancy of the nicotine concentrations between the blood levels and the levels needed for the *in vitro* chemokine production may be a lack of consideration of other signal(s), such as TCR [28] or IL-15 [29, 30]. IL-15 is expressed preferentially by non-lymphoid tissues, epithelial/fibroblast cell lines, and activated monocytes/macrophages, and induces IL-8 and RANTES expression in neutrophils and T cells, respectively. We observed that there were higher levels of IL-15 in smokers' blood, as compared with those in non-smokers' blood (unpublished observation). Therefore, the physiological concentration of nicotine may be enough *in vivo* for the upregulation of the chemokine production if synergized with IL-15. Precise mechanisms are under investigation in relation to the effective concentrations of nicotine and its responsible receptor subunit(s).

The activation of the NF-κB pathway seems to be a common route through which nicotine induces IL-8 and RANTES expression because these expressions were abrogated by the NF-κB inhibitors (IL-8 shown in [9]; RANTES not shown). However, the upstream events differed between neutrophils and T cells, i.e., the former requires the generation of ROIs, but the latter does not. From the candidate radicals $H^{\bullet}$, HOCl, and/or $ONOO^-$, $ONOO^-$ was shown to directly contribute to the activation of neutrophils, because NAC and L-NAME, but not DMSO, L-alanine, or L-methionine, completely inhibited all of the following successive events: DHR oxidation, IκB degradation, NF-κB activation, and IL-8 production. The $ONOO^-$ is thought to be formed by the reaction of nicotine-induced O_2^- with preformed NO, supported by the facts that constitutive NO was detected in unstimulated neutrophils, and its amount tended to decrease following nicotine stimulation. Regarding nicotine-stimulated T cells, the signals activating NF-κB are obscure.

Our study suggests that the blood levels of IL-8 and RANTES associate with heavy smoking which causes leukocytosis. Although further studies are required to prove the idea, it is presumable that chronic activation of neutrophils causes certain changes in the physiological status in smokers. One of the possibilities is that smoking may reduce the capacity of neutrophils to accomplish the specific immune responses, since intravenous IL-8 reduces neutrophil recruitment to sites of inflammation [31, 32]. RANTES activates and induces T cell proliferation. It also mediates degranulation of basophils and induces respiratory bursts in eosinophils. Furthermore, RANTES is associated with resistance to HIV. Based on these biological functions, control of RANTES gene expression has been thought to be useful for immunotherapies for a variety of diseases [33]. In this respect, it should be clarified how the chronic elevation of RANTES expression, as well as IL-8 expression, is accomplished in smokers, because this process may cause the disturbance in the immune homeostasis, and influence the appearance, treatment, and recovery of diseases.

CONCLUSION

1. Blood levels of IL-8, MCP-1, and RANTES were elevated in smokers.
2. Nicotine enhanced, *in vitro*, IL-8 production by neutrophils, and this enhancement seemed to be responsible for an increase in the number of neutrophils in smokers.
3. Nicotine enhanced, *in vitro*, RNATES production by T cells, and MCP-1 production by neutrophils. These chemokines may be responsible for lymphocytosis and monocytosis in smokers due to the action of mixed leukocyte recruitment.
4. The mechanism by which nicotine induces chemokine production was thought to be mediated by the activation of NF-κB. However, the upstream signal was different between neutrophils and T-cells; i.e., $ONOO^-$ is involved in the activation of neutrophils, but not of T cells.
5. Blood levels of IL-8 and RANTES appeared to be adequate markers for a criterion of heavy smoking as associated with leukocytosis.

Acknowledgments

This study was supported by the Smoking Research Foundation. We are grateful to Drs. S. Itani, Y. Nojyo, T. Muramatsu, H. Iwasakai, R. Takauji, H. Takatsuka, Y. Osawa, K. Nakamura, Y. Sasaki, and T. Matsuki, and Ms. Y. Tanaka, C. Kobayashi, and S. Horiguchi.

References

[1] Benowitz, N.L. (1986). Clinical pharmacology of nicotine. *Ann. Rev. Med. 37*, 21-32.

[2] Corre, F., Lellouch, J. & Schwartz, D. (1971). Smoking and leukocyte-counts. Results of an epidemiological survey. *Lancet,* **2**, 632-634.

[3] Schwartz, J. & Weiss, S.T. (1994). Cigarette smoking and peripheral blood leukocyte differentials. *Ann. Epidemiol.,* **4**, 236-242.

[4] Freedman, D.S., Flanders, W.D., Barboriak, J.J., Malarcher, A.M. & Gates, L. (1996). Cigarette smoking and leukocyte subpopulations in men. *Ann. Epidemiol.,* **6**, 299-306.

[5] Terashima, T., Wiggs, B., English, D., Hogg, J.C. & van Eeden, S.F. (1997). The effect of cigarette smoking on the bone marrow. *Am. J. Respir. Crit. Care Med.,* **155**, 1021-1026.

[6] Cassatella, M.A. (1999). Neutrophil-derived proteins: selling cytokines by the pound. *Adv. Immunol.,* **73**, 369-509.

[7] Larsen, C.G., Anderson, A.O., Appella, E., Oppenheim, J.J. & Matsushima, K. (1989). The neutrophil-activating protein (NAP-1) is also chemotactic for T lymphocytes. *Science,* **243**, 1464-1466.

[8] Taub, D.D., Anver, M., Oppenheim, J.J., Longo, D.L. & Murphy, W.J. (1996). T lymphocyte recruitment by interleukin-8 (IL-8). IL-8-induced degranulation of neutrophils releases potent chemoattractants for human T lymphocytes both in vitro and in vivo. *J. Clin. Invest.,* **97**, 1931-1941.

[9] Iho, S., Tanaka, Y., Takauji, R., Kobayashi, C., Muramatsu, I., Iwasaki, H., Nakamura, K., Sasaki, Y., Nakao, K. & Takahashi, T. (2003). Nicoine induces human neutrophils to produce IL-8 through the generation of peroxynitrite and subsequent activation of NF-κB. *J. Leukoc. Biol.,* **74**, 942-951.

[10] Davies, B.D., Hoss, W., Lin, J.P. & Lionetti, F. (1982). Evidence for a noncholinergic nicotine receptor on human phagocytic leukocytes. *Mol. Cell. Biochem.,* **44**, 23-31.

[11] Lebargy, F., Benhammou, K., Morin, D., Zini, R., Urien, S., Brée, F., Bignon, J., Branellec, A. & Lagrue, G. (1996). Tobacco smoking induces expression of very-high-affinity nicotine binding sites on blood polymorphonuclear cells. *Am. J. Respir. Crit. Care Med.,* **153**, 1056-1063.

[12] Hulea, S.A., Olinescu, R., Nita, S., Crocnan, D. & Kummerow, F.A. (1995). Cigarette smoking causes biochemical changes in blood that are suggestive of oxidative stress: a case-control study. *J. Environ. Pathol. Toxicol. Oncol.,* **14**, 173-180.

[13] van Antwerpen, L., Theron, A.J., Myer, M.S., Richards, G.A., Wolmarans, L., Booysen, U., van der Merwe, C.A., Sluis-Cremer, G.K. & Anderson, R. (1993). Cigarette smoke-mediated oxidant stress, phagocytes, vitamin C, vitamin E, and tissue injury. *Ann. N Y. Acad. Sci.,* **686**, 53-65.

[14] Filep, J.G., Beauchamp, M., Baron, C. & Paquette, Y. (1998). Peroxynitrite mediates IL-8 gene expression and production in lipopolysaccharide-stimulated human whole blood. *J. Immunol.,* **161**, 5656-5662.

[15] Terashima, T., English, D., Hogg, J.C. & van Eeden, S.F. (1998). Release of polymorphonuclear leukocytes from the bone marrow by interleukin-8. *Blood,* **92**, 1062-1069.

[16] Jagels, M.A. & Hugli, T.E. (1992). Neutrophil chemotactic factors promote leukocytosis. A common mechanism for cellular recruitment from bone marrow. *J. Immunol.,* **148**, 1119-1128.

[17] Mio, T., Romberger, D.J., Thompson, A.B., Robbins, R.A., Heires, A. & Rennard, S.I. (1997). Cigarette smoke induces interleukin-8 release from human bronchial epithelial cells. *Am. J. Respir. Crit. Care Med.,* **155**, 1770-1776.

[18] Nishikawa, M., Kakemizu, N., Ito, T., Kudo, M., Kaneko, T., Suzuki, M., Udaka, N., Ikeda, H. & Okubo, T. (1999). Superoxide mediates cigarette smoke-induced infiltration of neutrophils into the airways through nuclear factor-κB activation and IL-8 mRNA expression in guinea pigs *in vivo*. *Am. J. Respir. Cell Mol. Biol.,* **20**, 189-198.

[19] Mizgerd J.P. (2002). Molecular mechanisms of neutrophil recruitment elicited by bacteria in the lungs. *Semin. Immunol.,* **14**, 123-132.

[20] Totti N., 3rd, McCusker, K.T., Campbell, E.J., Griffin, G.L. & Senior, R.M. (1984). Nicotine is chemotactic for neutrophils and enhances neutrophil responsiveness to chemotactic peptides. *Science,* **223**, 169-171.

[21] Tsuyuoka, R., Takahashi, T., Sasaki, T., Taniguchi, Y., Fukumoto, M., Suzuki, A., Nakamura, K., Kobayashi, S., Kudo, T. & Nakao, T. (1994). Colony-stimulatng factor-producing tumours: production of granulocyte colony-stimulating factor and interleukin-6 is secondary to interlekin-1 production. *Eur. J. Cancer 30A*, **2130**-2136.

[22] Baggiolini, M., Dewald, B. & Moser, B. (1994). Interleukin-8 and related chemotactic cytokines -- CXC and CC chemokines. *Adv. Immunol.,* **55**, 97-179.

[23] Rollins, B.J. (1997). Chemokines. *Blood,* **90**, 909-928.

[24] Scapini, P., Lapinet-Vera, J.A., Gasperini, S., Calzetti, F., Bazzoni, F. & Cassatella, M.A. (2000). The neutrophil as a cellular source of chemokines. *Immunol. Rev.,* **177**, 195-203.

[25] Russell, M.A., Jarvis, M., Iyer, R. & Feyerabend, C. (1980). Relation of nicotine yield of cigarettes to blood nicotine concentrations in smokers. *Br. Med. J.,* **280**, 972-976.

[26] Seow, W.K., Thong, Y.H., Nelson, R.D., Macfarlane, G.D. & Herzberg, M.C. (1994). Nicotine-induced release of elastase and eicosanoids by human neutrophils. *Inflammation,* **18**, 119-127.

[27] Feyerabend, C., Higenbottam, T. & Russell, M.A. (1982). Nicotine concentrations in urine and saliva of smokers and non-smokers. *Br. Med. J.,* **284**, 1002-1004.

[28] Swanson, B.J., Murakami, M., Mitchell, T.C., Kappler, J. & Marrack, P. RANTES production by memory phenotype T cells is controlled by a posttranscriptional, TCR-dependent process.(2002). *Immunity,* **17**, 605-615.

[29] McDonald, P.P., Russo, M.P., Ferrini, S. & Cassatella, M.A. (1998). Interleukin-15 (IL-15) induces NF-κB activation and IL-8 production in human neutrophils. *Blood,* **92**, 4828-4835.

[30] Chen, J.P., Liao, N.S., Lai, S.L., Hsu, L., Mao, W.Y., Ku, M.C. & Liao, F. (2005). Reduced 2,4-dinitro-1-fluorobenzene-induced contact hypersensitivity response in IL-15 receptor alpha-deficient mice correlates with diminished CCL5/RANTES and CXCL10/IP-10 expression. *Eur. J. Immunol.,* **35**, 690-698.

[31] Hechtman, D.H., Cybulsky, M.I., Fuchs, H.J., Baker, J.B. & Gimbrone, M.A. Jr. (1991). Intravascular IL-8. Inhibitor of polymorphonuclear leukocyte accumulation at sites of acute inflammation. *J. Immunol.,* **147**, 883-892.

[32] Ley, K., Baker, J.B., Cybulsky, M.I., Gimbrone, M.A. Jr. & Luscinskas, F.W. (1993). Intravenous interleukin-8 inhibits granulocyte emigration from rabbit mesenteric venules without altering L-selectin expression or leukocyte rolling. *J. Immunol.,* **151**, 6347-6357.

[33] Toka, F.N., Pack, C.D. & Rouse B.T. (2004). Molecular adjuvants for mucosal immunity. *Immunol. Rev.*, **199**, 100-112.

In: Chemokine Research Trends
Editor: Leah R. Grinwald, pp. 89-105

ISBN: 978-60021-588-9

Chapter 6

TARGETING MIP-1α AND ITS RECEPTOR CCR5 IN RODENT MODELS OF RHEUMATOID ARTHRITIS

Chrong-Reen Wang[1,2,4*], Ai-Li Shiau[2,4] and Chao-Liang Wu[3,4]

[1]Departments of Internal Medicine
[2]Microbiology and Immunology
[3]Biochemistry and Molecular Biology
[4]Institutes of Basic Medical Sciences
National Cheng Kung University, College of Medicine, Tainan, Taiwan

ABSTRACT

Given the rapid progress being made in the understanding of the migration of leukocytes to inflammatory sites, research aiming to target proinflammatory chemokines and their receptors in patients with rheumatoid arthritis (RA) and animal models of RA has been actively pursued in the last few years. Antagonists of a number of chemokines and chemokine receptors are in phase 1 and 2 trials for RA. In previous studies the expressions of chemokine receptors CCR1 to 9, CXCR1 to 5, CX_3CR1, and XCR1 were examined in mononuclear cells (MNCs) of paired synovial fluid and blood samples from RA patients by quantitative reverse transcription polymerase chain reaction. Highly elevated levels of CCR5 and XCR1 in synovial MNCs were consistently demonstrated in all patients. Moreover, levels of the corresponding ligands, MIP-1α and lymphotactin, were increased in synovial fluid. Therefore, in this study we examined the involvement of CCR5 and XCR1 in the DBA/1 mouse model and MIP-1α in the Sprague-Dawley rat model of collagen-induced arthritis (CIA). Greater numbers of $CCR5^+$ T cells infiltrated more severe inflamed joints, with a concomitant increase in MIP-1α production. In addition, MNCs from more severe inflamed joints expressed higher levels of XCR1. Depletion of $CCR5^+$ cells significantly decreased type II collagen-induced T cell proliferation and MIP-1α production. Adenovirus-mediated gene therapy that suppressed the MIP-1α promoter activity significantly reduced MIP-1α levels in ankle joint extracts and suppressed CIA. Taken together, our results from animal experiments suggest a therapeutic potential of targeting MIP-1α and its receptor CCR5 in RA patients.

Key Words: rheumatoid arthritis, MIP-1α, CCR5, XCR1, collagen-induced arthritis

---

[*] E-mail address: wangcr@mail.ncku.edu.tw, Tel: 886-6-2353535, ext 5366; Fax: 886-6-2766175

INTRODUCTION

Human chemokines represent a structurally related family of chemotactic cytokines which act on distinct subsets of leukocytes via their specific G protein-coupled cognate receptors expressed on the cell surface [1,2]. They consist of 70 to 130 amino acids with four conserved cysteines and two main subfamilies are distinguished according to the position of first two cysteines. The CXC chemokines are characterized by two cysteines separated by one amino acid, and the CC chemokines are identified by nearby cysteines. In addition, two variants of the chemokine structure are found, including lymphotactin with only two conserved cysteines (C) and fractalkine with three amino acids between the first two cysteines (CX_3C). To date, there are ten CC receptors (CCR1-10), six CXC receptors (CXCR1-6), one C receptor (XCR1) and one CX_3C receptor (CX_3CR1). Chemokines and their receptors are known to direct the trafficking of leukocytes through the body, participate the immune responses and contribute to the pathogenesis of human diseases such as autoimmunity, virus infection, allergy, cancer metastasis, atherosclerosis, allograft rejection and pulmonary inflammation [1,3-9].

Rheumatoid arthritis (RA) is a chronic autoimmune disease with synovial tissues invaded by activated mononuclear cells (MNCs) [10]. These cells and associated products such as chemokines play an essential role in synovitis, pannus formation and, eventually, joint destruction. Both synovial lining cells and infiltrating leukocytes are sources of these chemokines [11]. Chemokine receptors allow leukocytes to sense chemokines gradients, thereby directing these cells into joint compartments. As a result, a battery of secreted chemokines and their interaction with chemokine receptors accounts for the recruitment of the MNCs responsible for initiating and maintaining the immune responses within rheumatoid joint [11,12]. A number of studies aim at elucidating the role of chemokines in RA. Synovial fluids and synovial tissues of RA patients contain members of chemokines in the CC family, including monocyte chemoattractant protein-1 (MCP-1), macrophage inflammatory protein-1α (MIP-1α), MIP-3α and regulated upon activation normal T expressed and secreted (RANTES) [3,11-13]. They also contain chemokines in the CXC family, such as IL-8, epithelial-neutrophil activating protein-78 (ENA-78), growth-related gene product-α (Groα), connective tissue activating protein-III (CTAP-III), IFNγ inducible protein-10 (IP-10), platelet factor 4 (PF4), monokine induced by IFNγ (MIG) and stromal cell-derived factor-1 (SDF-1). Fractalkine, a CX_3C member family, and lympho-tactin, a C member family, have been demonstrated in rheumatoid joint [14,15]. Chemokine receptors have been found to be up-regulated in RA and recent studies suggest CCR1, CCR2, CCR5, CCR6, CXCR1, CXCR2, CXCR3, CXCR4, CX_3CR1 and XCR1 are critical chemokine receptors [3,12].

In previous studies, the status of chemokine receptor expression in RA patients was evaluated by the quantitative reverse transcription polymerase chain reaction (RT-PCR) [15]. The MNCs from paired synovial fluid and blood samples were studied for expressions of CCR1 to 9, CXCR1 to 5, CX_3CR1 and XCR1. Higher synovial expression levels of CCR5 and XCR1 as compared with paired blood samples were consistently demonstrated in all RA patients. Significant CCR5 display was found on infiltrating MNCs in rheumatoid joint by immunohistochemical staining, and XCR1 expression was detected in MNCs and synoviocytes in inflamed synovium by *in situ* hybridization [13,15]. Moreover, in each patient, the MIP-1α and lymphotactin levels of joint fluid were consistently higher than those of paired serum samples.

To test the hypothesis that targeting the chemokines/chemokine receptors could be an effective novel strategy for amelioration of arthritis in RA patients, in this study, we examined the functions of CCR5 and XCR1 in the DBA/1 mouse collagen-induced arthritis (CIA) and MIP-1α in the Sprague-Dawley (SD) rat CIA. Moreover, we depleted $CCR5^+$ cells in the DBA/1 mouse model by intraperitoneal (i.p.) injection of anti-CCR5 monoclonal antibody (mAb) and inhibited MIP-1α production in the SD rat model by intra-articular (i.a.) injection of adenovirus-mediated gene transfer that suppressed the promoter activity of MIP-1α.

MATERIALS AND METHODS

Animal Experiments

The animal studies were approved by the Laboratory Animal Care and Use Committee of Medical College, National Cheng Kung University. Both local and international guidelines for the humane treatment of laboratory animals were followed. For each experiment, animals were allocated into groups by randomization. These animals were housed under specific pathogen-free condition, fed a standard laboratory diet (PMI Nutrition International, St. Louis, MO) and given sterilized water *ad libitum*.

Induction of CIA in DBA/1 Mice and Evaluation of Arthritis

Eight-week old DBA/1(J) mice were purchased from Charles River Laboratory Japan (Kanagawa, Japan). Bovine type II collagen (Sigma, St. Louis, MO) was emulsified with an equal volume of complete Freund's adjuvant (CFA) containing 4 mg/ml of *Mycobacterium tuberculosis* (Chondrex, Redmond, WA). One hundred μl emulsion was injected subcutaneously (s.c.) into the tail base. Three weeks after immunization, mice were examined every two days. The presence of arthritis of each paw was graded 0 to 2 by a previously described score system, where 0 = normal, 1 = low score and 2 = high score [16].

Preparation of Mouse MNCs

After the development of arthritis (6 weeks after the immunization), the inflamed joint tissues with low or high score were removed from paws by first stripping the skin and separating the limb below the wrist or ankle joints [16]. The minced tissues were further subjected to 10 mg/ml collagenase (Sigma) treatment for 1 hr at 37^0C. After being filtrated through nylon mesh, these cells were subjected to centrifuge over Lympholyte-M (Cedarlane, Hornby, Canada) to purify MNCs. There were no available MNCs from joints of normal score. Spleens and regional lymph nodes near arthritic paws including axillary and popliteal sites were separately harvested and were teased apart to make single-cell suspension. Red blood cells were depleted by a mouse erythrocyte lysing kit (R&D System, Minneapolis, MN).

Flow Cytometry Analysis of Mouse MNCs

Double fluorescence staining was performed for the surface phenotype analysis after the Fc receptors of MNCs were blocked with anti-CD16/CD32 (clone 2.4G2, BD PharMingen, San Diego, CA) [16]. The mAbs against mouse antigens included fluorescein isothiocyanate (FITC)-conjugated anti-CD3 (clone 145-2C11, BD PharMingen), FITC-conjugated anti-CD4 (clone GK1.5, BD PharMingen) or FITC-conjugated anti-CD8 (clone 53-6.7, BD PharMingen) and phycoerythrin (PE)-conjugated anti-CCR5 (clone C34-3448, BD PharMingen) or PE-conjugated anti-CCR3 (clone 83101.111, R&D System). After washing, the stained cells were analyzed with a FACSort and CellQuest software programs (Becton Dickison, Mountain View, CA). The stainings of control isotype mAbs (BD PharMingen) were included for each sample.

RT-PCR of CCR5 and XCR1

Total RNA was extracted from MNCs by a PUREscrpit RNA isolation kit (Gentra, Minneapolis, MN). RT-PCR was performed with 1 μg of RNA, 20 pmol of oligo $(dT)_{18}$ primer, 0.5 mM dNTP, 0.5 U of RNase inhibitor, and 200 U of MLV RTase in 4 μl of 5X RT buffer. All reagents used in the cDNA synthesis were purchased from BD Clontech (Palo Alto, CA). The mixture was incubated at 42^0C for 1 hr and 95^0C for 5 min. For mouse CCR5, PCR analysis of full-length CCR5 mRNA (1,078 bp) expression was done with 1 cycle of 94^0C for 5 min and a total of 40 cycles of 94^0C for 1 min, 55^0C for 1 min and 72^0C for 2 min using forward primer (5'-GGCTCTTGCAGGATGGATCT-3') and reverse primer (5'-GTCATAAACCAGTAGAAAC-3') [17]. For mouse XCR1, the XCR1 mRNA (431 bp) expression was done with 1 cycle of 94^0C for 5 min and a total of 40 cycles of 94^0C for 1 min, 54^0C for 1 min and 72^0C for 1 min using forward primer (5'-CTCCTGTCTACTG CCTGTGTTG-3') and reverse primer (5'-TGACTGTTCGGTGTCTCTGTCT-3') [18]. The mouse housekeeping gene hypoxanthine phosphoribosyl transferase (HPRT) was also amplified using forward primer (5'-GTTGGATACAGGCCAGACTTTGTTG-3') and reverse primer (5'-GAGGGTAGGCTGGCCTATAGGCT-3'). All of the PCR products were resolved using ethidium bromide-stained 1% agarose gels.

Purification of $CD4^+$T Cells and Depletion of $CCR5^+$Cells

The MNCs were stained with rat anti-mouse CD4 mAb (clone GK1.5, BD PharMingen), rat anti-mouse CCR5 mAb (clone C34-3448, BD PharMingen) or control rat IgG (Sigma), and were further incubated with magnetic particles coated anti-IgG microbeads (Miltenyi Biotec, Germany). These cells were further passed through a positive selection column ($CD4^+$ cells) or a negative selection column ($CCR5^+$ cells) in a Magnetic Cell Sorter (Miltenyi Biotec). The column was removed from the separator and the positive fraction of $CD4^+$T cells was flushed out with purity more than 95%. The contamination of CCR5$^+$cells in the preparation of negative selection was less than 5%.

Quantification of MIP-1α and IFN-γ

MIP-1α and IFN-γ levels were quantified by enzyme-linked immunosorbent assay (ELISA) kits (R&D System) according to the manufacteur's instruction. The MNCs in the density of 4 x 10^5 per well in 96-well flat-bottomed microplates were cultured with 100 μg/ml of denatured type II collagen in RPMI 1640 medium supplemented with 10% heat-inactivated fetal calf serum, 2 mM glutamine, and antibiotics. CD4$^+$T cells in the density of 3 x 10^5 per well in 96-well round- bottomed microplates were cultured. Culture supernatants were collected serially at 24, 48 and 72 hr and the contents of MIP-1α and IFN-γ were measured. The inflamed joints were removed from paws and synovial tissues were minced in cold phosphate-buffered saline (PBS) and incubated on ice for 1 hr to release soluble MIP-1α and IFN-γ.

Cell Proliferation Assay

The MNCs in the density of 4 x 10^5 per well in 96-well flat-bottomed microplates were cultured with 10, 30 or 100 μg/ml of denatured type II collagen in RPMI 1640 medium supplemented with 10% heat-inactivated fetal calf serum, 2 mM glutamine, and antibiotics. Cell proliferation was measured by pulsing cells for the last 10 hr of 72- hr culture with 1 μCi/well of ^{3}[H] thymidine.

In Vivo Depletion of CCR5$^+$Cells in Mouse CIA

The *in vivo* experiments to deplete CCR5$^+$cells and to assess this effect on CIA were performed. Mice were injected i.p. with 0.5 mg rat anti-mouse CCR5 mAb (clone C34-3448, BD PharMingen, San Diego, CA) for the experiment group or 0.5 mg rat IgG (Sigma) for the control group. Five mice per group were injected with mAb or IgG before priming with type II collagen in CFA. These mice were recorded for their clinical scores every 2 days up to 12 weeks after immunization for assessment of the therapeutic effect.

Generation of Adenoviral Vectors

The adenoviral vector encoding nuclear localization signal-deleted prothymosin (Ad-ProTdN), was generated as previously described [19]. The AdLacZ, an adenoviral vector encoding-galactosidase, was used as the control vector. Recombinant viruses were propagated in 293 cells and viral titer was measured by the plaque assay using 293 cells.

Assay of MIP-1α Promoter in Cells Following AdProTdN Gene Transfer

Mouse MC3T3-E1 osteoblasts grown in 24-well plates were transfected with 0.8 μg of the pGL3 luciferase reporter plasmid with human MIP-1α promoter, which was kindly provided by Dr. P. Shore at the University of Manchester [20], by lipofectamine 2000 (Invitrogen, Carlsbad, CA). Four hours after transfection, the cells were stimulated with 500 ng/ml of lipopolysaccharide (LPS) for 16 hr followed by infection with AdProTdN or AdLacZ at multiplicities of infection (MOI) of 10. Cell lysates were harvested after 48 hr and the luciferase activity was measured by the Bright-Gro luciferase assay system (Promega, Madison, WI). The luciferase activity was expressed as relative light units (RLU)/μg protein.

ELISA for MIP-1α in Rat Joint Extracts

Synovial tissues taken from the ankle joints of CIA rats on day 27 were snap frozen and homogenized. The concentrations of MIP-1α in the ankle homogenates were quantified using ELISA kits (R&D system) in accordance with the manufacturer's instructions. Total protein concentrations were determined using a BCA protein assay kit (Pierce, Rockford, IL).

Chemotaxis Assay of AdProTdN Gene Transfer

The chemotaxis assay was preformed using a 48-well modified Boyden chamber (Neuroprobe, Cabinjohn, MD) with 8-μm pore polycarbonate filters (Neuroprobe). MC3T3-E1 cells were infected with AdProTdN or AdLacZ at an MOI of 10 and the conditioned media were collected 48 hr later. In addition, the supernatant from MC3T3-E1 cells stimulated with LPS (1 μg/ml) for 16 hr was used as the positive control. The RAW 264.7 macrophage-like cells ($5x10^4$) were added to the upper chamber and the conditioned medium was added to the lower chamber. After incubation for 5 hr, the cells attached to the lower filter surface of the membrane were fixed by methanol, stained with toluidine blue, and counted in a microscope.

Induction of CIA in SD Rats and Evaluation of Arthritis

SD rats (male, 6 to 7 weeks of age) were obtained from the Laboratory Animal Center in Medical College of National Cheng Kung University. Rats were immunized on day 0 with intradermal (i.d.) injection at four sites on the back of 0.4 ml 1:1 mixture of 400 μg bovine type II collagen (Elastin Products, Owensville, MO) in 0.1 M acetic acid and CFA (Sigma) [21]. The booster doses were injected i.d. on day 7 with 0.2 ml of the same emulsion into posterior aspects of both hip joints. The ankle circumferences were determined after the injection of booster doses by a dial caliper with a geometric formula, $2\pi(\sqrt{(a^2 + b^2 / 2)})$,where a is the latero-lateral diameter and b is the antero-posterior diameter.

Protocol of AdProTdN Gene Therapy

For administration of viruses into the joints, both ankles of rats were injected on days 7, 12, 17 and 22 with 5 x 10^7 plaque-forming units of AdLacZ or AdProTdN, each group with 20 ankles. The ankle joints were held in 15 to 20 degrees of plantar flexion and a 31-gauge needles on a plastic syringe was inserted antero-laterally at the ankle joint line about 1 to 2 mm distal to the lateral malleolus. The ankle circumferences were measured by a dial caliper on days 1, 7, 12, 17, 22 and 27.

Histopathological Assessment and Immuno- Histochemical Staining of Macrophages

Ankles were resected and fixed in formalin. After further decalcification, ankle joints were embedded in paraffin, sectioned, and stained with hematoxylin and eosin. The presence of synovial hyperplasia, cartilage erosion, and leukocytes infiltrates was evaluated and scored on a scale of 0 to 2 for each item (0 = absent, 1 = mild, 2 = severe) [21]. The synovial tissues from ankle joints were embedded in OCT compound and 5 μm cryostatic sections were prepared. The frozen sections were stained with mouse anti-rat ED-1 (Chemicon, Temecula, CA) mAb or mouse IgG as the control Ab. The sections were further incubated with biotinylated anti-mouse IgG and detected with streptavidin/peroxidase staining. The diamino-benzidine was used as the substrate chromogen. The slides were counterstained with hematoxylin. Quantification of macrophage numbers was done by the image analysis software (Image-Pro Plus, Media Cybernetics, Silver Spring, MD).

Statistical Analysis

Data were expressed as mean $\pm$ standard error of the mean (SEM). Differences between clinical scores of anti-CCR5 mAb and control IgG-treated groups, and ankle circumferences of AdProTdN- and AdLacZ-treated groups were compared by two-way ANOVA with Bonferroni adjustment for multiple comparisons. Other data obtained from two groups were compared with Student's *t* test. The *p* values less than 0.05 were considered significant.

RESULTS

Infiltration of $CCR5^+$T Cells into Mouse Inflamed Joints

As shown on Figure 1A, the mRNA expression of CCR5 was higher in MNCs from inflamed joints than regional lymph nodes (LN) and spleens (Spl) in the mouse CIA. The intensity of CCR5 was much stronger in inflamed joints of high score as compared with low score (data not shown). The infiltrating synovial $CCR5^+$T cells were shown on Figure 1B. Most of these cells were $CD4^+$T cells with less than 10 % of $CD8^+$T cells. The percentages of CCR5 expression on $CD4^+$T cells were higher in inflamed joints of high score than low score (14.7 ± 2.2 % *vs*. 7.8 ± 2.8 %, $p < 0.05$, data from 3 experiments). There was no expression of

CCR3 (a marker of Th2 cells) on synovial CD4$^+$T cells, which suggested that most of infiltrating T cells belong to the Th1 subset.

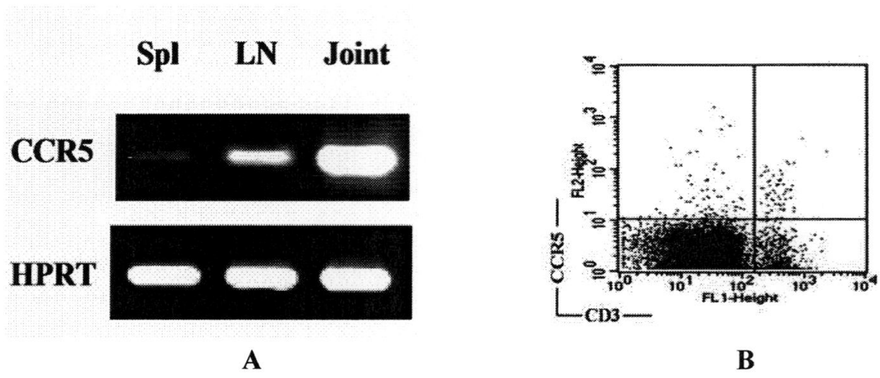

Figure 1.

Production of MIP-1α by Synovial CCR5$^+$T Cells

The MIP-1α levels were significantly higher in joint extracts from paws of high score as compared with low score (Figure 2A, 30.6 ± 5.8 *vs*. 10.2 ± 0.5, pg per paw, $p < 0.005$, data from 12 paws per group). In addition, higher levels of IFN-γ were found in joint extracts from paws of high score as compared with low score (25.2 ± 9.8 *vs*. 12.7 ± 4.9 pg per paw, $p < 0.05$, data from 12 paws per group). High amounts of MIP-1α and IFN-γ were produced by *in vitro* culture of CD4$^+$ T cells from joints (at 72 hr, for MIP-1α, 206.7 ± 20.2 pg/ml, for IFN-γ, 52.0 ± 9.5 pg/ml). These data were consistent with previous reports that MIP-1α could be co-secreted with IFNγ by Th1 cells [22]. As shown on Figure 2B, the depletion of CCR5$^+$ cells from MNCs significantly decreased the *in vitro* MIP-1α production (at 72 hr, CCR5 depletion *vs*. control IgG, 186.2 ± 22.3 *vs*. 246.1 ± 20.5 pg/ml, $p < 0.05$, data from 3 experiments).

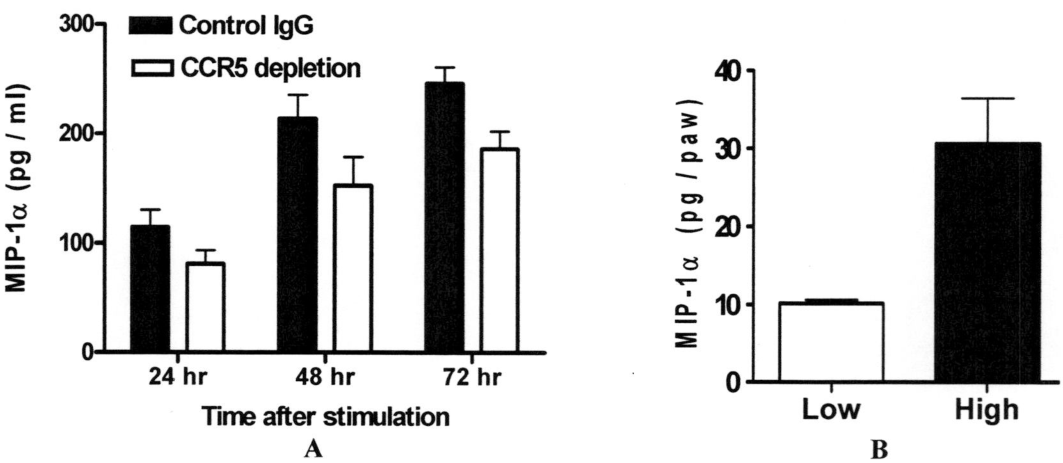

Figure 2.

In vitro and in vivo Depletion of CCR5$^+$ Cells in the Mouse CIA

Depletion of CCR5$^+$cells significantly decreased the IFN-γ production by *in vitro* culture of MNCs (at 72 hr, CCR5 depletion *vs.* control IgG, 87.1 $\pm$ 37.5 *vs.* 219.8 $\pm$ 36.6 pg/ml, $p <$ 0.05, data from 3 experiments, Figure 3A). The kinetic profiles of MNCs proliferation by the addition of type II collagen were demonstrated in Figure 3B, and the *in vitro* depletion of the CCR5$^+$ portion significantly decreased the proliferation (at 72 hr, CCR5 depletion *vs.* control IgG, 9.1 $\pm$ 0.5 *vs.* 10.3 $\pm$ 0.5, $p < 0.05$). However, the *in vivo* depletion experiment using CCR5 mAb failed to show significant difference in clinical scores between two groups of mice (data not shown).

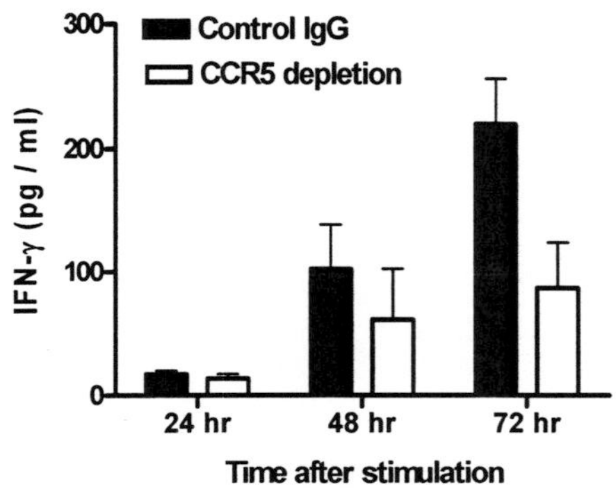

A

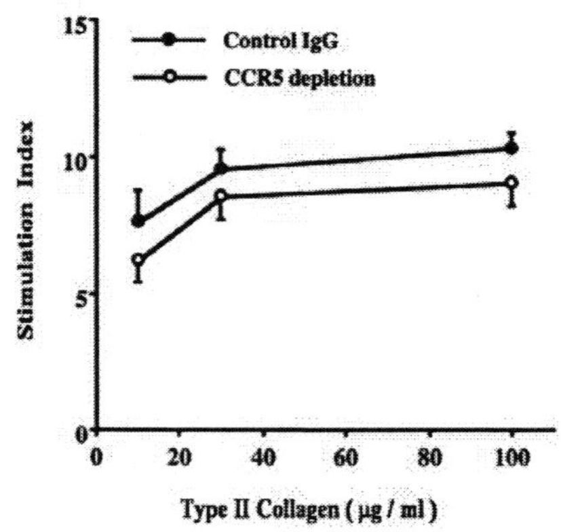

B

Figure 3.

Expression of XCR1 in the Mouse CIA

The RT-PCR of XCR1 was performed in MNCs form DBA/1 mice after the development of arthritis (6 weeks after the immunization). There was an up-regulation of XCR1 in MNCs from arthritic joints comparing to those in spleens (data not shown) and regional lymph nodes (LN) (Figure 4). The MNCs from high score inflamed joints expressed higher levels of XCR1 as compared with low score inflamed joints.

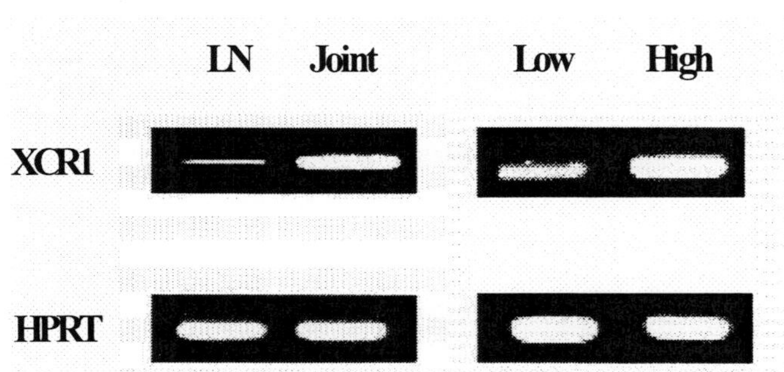

Figure 4.

Suppression of MIP-1α Promoter Activity and Macrophage Chemotaxis by AdProTdN Gene Transfer

The ProT, a key regulator in the cellular proliferation, has many immunomodulatory functions and deletion of its nuclear localization signal could abolish the growth-promoting capacity [23]. The fact that MIP-1α plays an essential role in the selective recruitment of macrophages into rheumatoid synovitis incited us to examine whether the AdProTdN gene transfer down-regulated the MIP-1α promoter activity. As shown in Figure 5 A, the MIP-1α promoter activity was augmented by LPS stimulation. Furthermore, the AdProTdN gene transfer significantly suppressed the MIP-1α promoter activity (AdProTdN *vs.* AdLacZ, 234,884 ± 23,055 *vs.* 424,474 ± 15,732, $p < 0.001$). To establish whether the AdProTdN transfer affected migration of macrophages, a chemotaxis assay was performed (Figure 5B). Although the conditioned media from AdLacZ-treated MC3T3-E1 cells triggered an increase in migration comparing with the basal levels, those from AdProTdN-treated cells significantly decreased the migrating number of macrophages (142.6 ± 32.1 *vs.* 317.6 ± 20.5, $p < 0.001$). These critical findings encouraged further studies applying the AdProTdN gene therapy into the rat model of CIA.

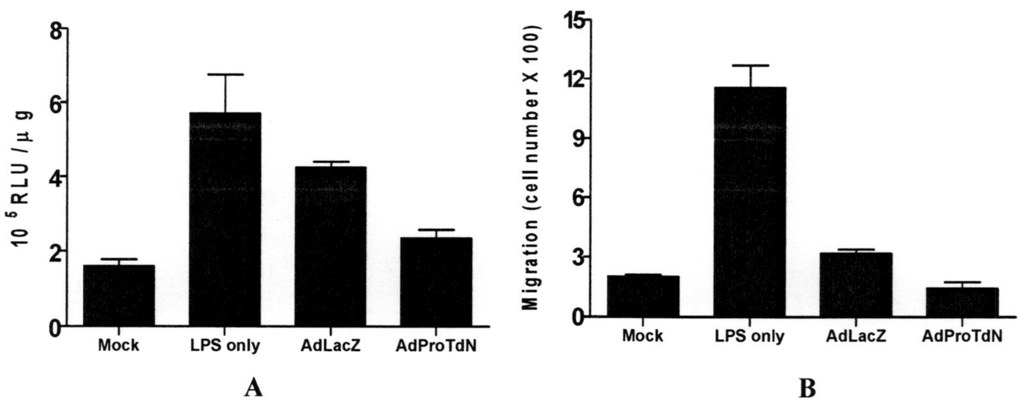

Figure 5.

Reduction of CIA and Inhibition of MIP-1α Production by AdProTdN Gene Therapy

There were statistically significant decreases of ankle circumferences in AdProTdN-treated joints as compared with the AdLacZ-injected ankles (Figure 6 A, $p < 0.001$ on days 22 and 27). The AdProTdN-treated joints had significantly reduced MIP-1α levels (normalized to total protein levels) as compared with AdLacZ-injected ankles (Figure 6B, 41.7 ± 3.6 *vs.* 55.2 ± 1.1 pg/mg protein, $p < 0.05$).

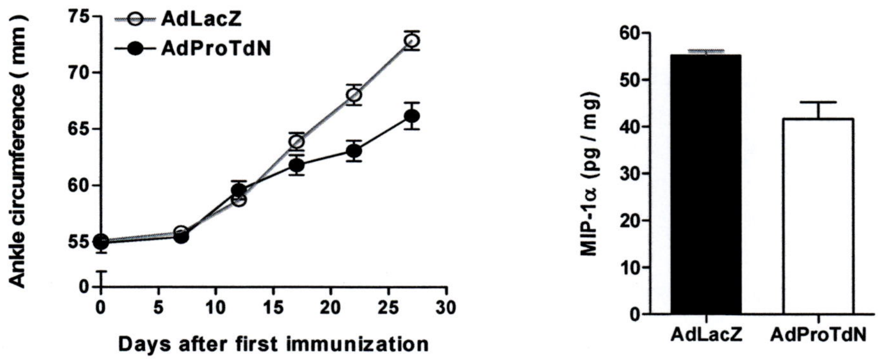

A B

Figure 6.

Repressed Macrophage Infiltration Into Ankle Joints by AdProTdN Gene Therapy

The clinical findings were further confirmed with histological evaluations. The AdProTdN-treated joints showed significant reductions of leukocyte infiltrates, synovial hyperplasia and bone erosion (Figure 7A, original magnification × 100). Histological scores were significantly lower in AdProTdN-injected ankles as compared with the AdLacZ-treated joints (1.50 ± 0.20 vs. 3.70 ± 0.68, $p < 0.05$). Moreover, the ED1 immunohistochemical staining was performed for identification of macrophages. The macrophage numbers were significantly lower in the AdProTdN-injected ankles comparing with the AdLacZ-treated joints (Figure 7B, 1.38 ± 0.15 vs. 2.76 ± 0.35, $p < 0.01$).

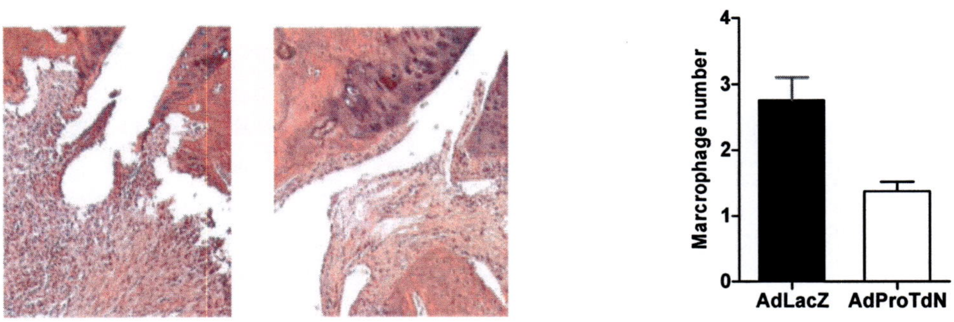

AdLacZ AdProTdN

A B

Figure 7.

DISCUSSION

Chemokine receptor CCR5 is used by the R5 human immunodeficiency virus-1 (HIV-1) strains as a coreceptor in the early stage of infection [4]. Mutations at the CCR5 promoter down-regulate the CCR5 expression at the transcription levels and further decrease its surface expression levels [24,25]. A nucleotide transition from C to T at position 59653 in the CCR5 promoter has been shown to slow the progression of HIV infection, and a polymorphic change at position 59029 in the CCR5 promoter has also been shown to affect the rate of progression to acquired immunodeficiency syndrome [24,25]. A 32-nucleotide deletion in CCR5 coding region produces a truncated protein that fails to reach the cell surface and, consequently, homozygous individuals with this deletion resist HIV infection [26]. Both *in vitro* and *in vivo* findings have also suggested that the CCR5 ligands, including RANTES, MIP-1α and MIP-1β, are inhibitors of cell entry and replication of HIV-1 [26,27]. The contribution of human leukocyte antigen class II alleles to RA has been estimated at less than 50% of the total genetic component, and the additional genetic contribution is necessary [28]. Further analysis of population genetics provides additional support to the hypothesis that the chemokine signaling through the CCR5 receptor contributes to the recruitment of leukocytes and thus to the development of RA. A study found that homozygous knockout existed in 7 out of 815 Caucasian controls, whereas none was homozygous for a CCR5 32-nucleotide deletion polymorphism in 673 RA patients that rendered CCR5 receptor inactive [29]. In addition, an increased frequency of CCR5 promoter 59653 T homozygous genotype has been found in RA patients [13]. There is increased evidence suggesting the role of chemokines in RA pathogenesis [11,12]. The polymorphism in the promoter region of chemokine gene has been reported to be associated with RA susceptibility. Patients with RA in Taiwan had significantly higher frequencies of the A allele and A/A genotype at the promoter -403 position of RANTES [30]. The relative contribution made by CCR5 and its ligands to the development of RA remains to be determined. As shown in previous reports, abundant amounts of MIP-1 α and RANTES and their corresponding receptor CCR5 have been detected in rheumatoid joints and their expressions are associated with the disease progression of RA [11,12]. Drugs preventing the entry of HIV into host cells by blocking CCR5, either with small molecular antagonists or with mAbs, are in rapid clinical development [1,4]. These facts prompt us to the idea that targeting CCR5 or its ligands such as MIP-1 α might be beneficial in RA patients.

Although the expression of a given chemokine receptor does not necessarily imply its involvement in the progression of disease, increased numbers of highly activated Th1 cells predominantly expressing CCR5 have been found in rheumatoid joints [13,15]. The infiltration of $CCR5^{+}CD4^{+}$T cells could be regulated by the synovial microenvironment abounding with IL-15. In order to investigate the feasibility of therapeutic manipulation of CCR5 molecules in RA patients, first we examined the function of $CCR5^{+}$T cells in the DBA/1 CIA mouse model. Most of the infiltrating $CCR5^{+}$T cells in joints were $CD4^{+}$T cells. The CCR5 expression was up-regulated on T cells from joints and greater numbers of $CCR5^{+}$ T cells infiltrated more severe inflamed joints. Lack of CCR3 (a Th2 marker) expression and production of IFN-γ upon activation suggested that $CCR5^{+}CD4^{+}$T cells belong to the Th1 subset. Th1-type immune responses are essential for the pathogenesis of organ-specific autoimmunity such as RA [31]. In the CIA model, Th1 cells play a pro-inflammatory role, while Th2 cells have an anti-inflammatory effect [32]. These findings are consistent with the results of this study that depletion of $CCR5^+$ cells significantly reduced the autoantigen-specific T cell proliferation and the concomitant IFN-γ production. Further *in vivo* depletion of CCR5 failed to demonstrate the therapeutic benefit in amelioration of arthritis. The fact that treating wild type mice with anti-CCR5 mAb does not affect the severity of CIA, as shown in our data and another study, suggests that CCR5 might not be a critical determinant in CIA [33]. However, the optimal dosages and schedules of anti-CCR5 mAb therapy remain to be determined in further investigation. There are a number of amino terminus-modified chemokine analogues that bind CCR5 and promote its internalization [34,35]. Indeed, the severity of the mouse model of CIA could be reduced by administering Met-RANTES through the inhibition of chemokines binding to CCR5 [35]. Another method to suppress the function of chemokine receptor is to employ small molecules to block ligand-bound or ligand-induced signaling [36,37]. In the CIA models of rat and monkey, chemical antagonists of CCR5 lead to protection against arthritis. In addition, the CCR5 knockout mice are protected in the incidence of CIA [38]. The anti-arthritis potential of CCR5 antagonists in CIA may provide a rationale for further clinical evaluation of the CCR5-targeted therapy in RA.

The only C family chemokine receptor XCR1 is expressed in spleen, thymus and circulating leukocytes including T and B lymphocytes and neutrophils [39]. The XCR1 expression has also been found in synoviocytes and MNCs within rheumatoid joints [15,40]. Although the expression of lymphotactin is an integral part in the differentiation of Th1 cells and in the pathogenesis of Th1-related autoimmunity, the relative importance of its receptor XCR1 in the Th1-mediated immune responses remains obscure [41]. In the present study, there was an up-regulation of XCR1 in MNCs from arthritic tissues and higher levels of XCR1 were expressed in more severe inflamed joints. Further investigations are needed to understand the mechanisms underlying the regulation of XCR1 expression within rheumatoid joints.

The MIP-1 α plays a crucial role in the recruitment of leukocytes into inflamed joints [11]. An abundant amount of MIP-1α has been found in synovial fluid of RA patients, and MIP-1α has been shown to account for one third of synovial fluid-derived chemotactic activity for MNCs [11,42]. In addition, MIP-1α has been shown to be derived from activated $CD4^+$T cells and fibroblasts from rheumatoid joints [13]. Increased MIP-1α production has been observed in the mouse CIA and rat adjuvant-induced arthritis [43,44]. Moreover, in this study, higher amount of MIP-1α was found in more severe inflamed joints and depletion of $CCR5^+$cells significantly reduced MIP-1α production. The relevance of these findings on the functional role of MIP-1α is clearly established in rodent models of CIA and RA patients. Therefore, targeting MIP-1α is worthy of consideration in arthritis alleviation. Indeed, anti-MIP-1α mAb treatment delayed the onset and reduced the severity of CIA and inhibited the development of arthritis in anti-type II collagen antibody-induced arthritis [43,45].

The unique characteristics of gene therapy include the ability to deliver therapeutic gene products to specific cells or tissues in a targeted, sustained, regulated and cost-effective fashion and the potentiality of bypassing the inherited obstacles associated with the transfer of therapeutic compounds [46]. Considerable progress has been made in the preclinical development of gene therapy for animal models of arthritis [46,47]. The original concept for the arthritis gene therapy is delivery of genes to individual diseased joints and local gene transfer can achieve sustained therapeutic concentrations of transgene products within the joint. Biological agents modulating the pro-inflammatory activities of TNF-α and IL-1β have shown efficacy as novel anti-arthritic drugs in RA; however, these therapies are limited by side effects such as infection as well as their excessive expense [48]. Furthermore, the arthritis activity relapses upon cessation of these drugs. There is a clear need for novel treatments in RA. As a potential clinical therapy, gene transfer could be viewed within the context of biological agents. Several phase 1 trials in gene therapy of RA patients have been successfully conducted and phase 2 trials are underway [46,47]. In this study, a significant down-regulation in luciferase activity of MIP-1α promoter was found in cells infected with AdProTdN. By using the *in vitro* chemotaxis assay, AdProTdN treatment resulted in a significant decrease in the migration of macrophages. Reduced *in vivo* MIP-1α concentrations of joint extracts was also noted in ankles injected with AdProTdN. Moreover, the *in vivo* macrophage infiltration into ankle joints was markedly suppressed by the AdProTdN gene therapy. Taken together, these data clearly demonstrated a possibility that the therapeutic effect of AdProTdN on arthritis is mediated through the inhibition of MIP-1α–induced recruitment of macrophages. This study is the first to exploit the strategy of down-regulation of MIP-1α promoter to abate arthritis in a mouse CIA model.

In conclusion, our data demonstrate the feasibility of arthritis amelioration by targeting the MIP-1 α and its receptor CCR5 in the rodent CIA models. Despite the redundancy in the system of chemokines/chemokine receptors, blocking of certain chemokines or chemokine receptors might be of value in the treatment of RA patients. No beneficial effect on disease activity was observed in RA patients receiving anti-MCP-1 mAb therapy [49]. A phase 1 trial using oral CCR1 small molecule antagonist resulted in a striking reduction of synovitis in RA patients [50]. This success encourages further clinical trials aiming at other members of chemokines/chemokine receptors repertoire. At the present time, antagonists of CCR1 and CCR2 are in phase 1 and 2 clinical trials [1]. Thus, targeting CCR5 or its ligand MIP-1α, either alone or adjunctively, holds a potential in RA therapy.

Acknowledgments

This work was supported by grants NSC 89-2314-B-006-098, 89-2314-B-006-142, 90-2314-B-006-094, 91-2314-B-006-019, 92-2314-B-006-047, 93-2314-B-006-010, 94-2314-B-006-052 and 95-2314-B-006-010 from the National Science Council of Taiwan.

References

[1] Charo IF, Ransohoff RM. The many roles of chemokines and chemokine receptors in inflammation. *N Engl J Med* 2006; 354:610-621.

[2] Cyster JG. Chemokines, sphingosine-1-phosphate, and cell migration in secondary lymphoid organs. *Annu Rev Immunol* 2005; 23:127-159.

[3] Koch AE. Chemokines and their receptors in rheumatoid arthritis: future targets? *Arthritis Rheum* 2005; 52:710-721.

[4] Lederman MM, Penn-Nicholson A, Cho M, Mosier D. Biology of CCR5 and its role in HIV infection and treatment. *JAMA* 2006; 296:815-826.

[5] Wang CR, Chen PC, Liu MF. Increased circulating CCR3+ type 2 helper T cells in house dust mite-sensitive Chinese patients with allergic diseases. *Asian Pac J Allergy Immunol* 2003; 21: 205-210.

[6] Yan L, Anderson GM, DeWitte M, Nakada MT. Therapeutic potential of cytokine and chemokine antagonists in cancer therapy. *Eur J Cancer* 2006; 42:793-802.

[7] Sheikine YA, Hansson GK. Chemokines as potential therapeutic targets in atherosclerosis. *Curr Drug Targets* 2006; 7:13-27.

[8] Fischereder M, Luckow B, Hocher B, et al. CC chemokine receptor 5 and renal transplant survival. *Lancet* 2001; 357: 1758-1761.

[9] Pignatti P, Brunetti G, Moretto D, et al. Role of the chemokine receptors CXCR3 and CCR4 in human pulmonary fibrosis. *Am J Respir Crit Care Med* 2006; 173:310-317.

[10] Firestein GS. Evolving concepts of rheumatoid arthritis. *Nature* 2003; 423:356-361.

[11] Szekanecz Z, Strieter RM, Kunkel SL, Koch AE. Chemokines in rheumatoid arthritis. *Springer Semin Immunopathol* 1998; 20:115-132.

[12] Szekanecz Z, Szucs G, Szanto S, Koch AE. Chemokines in rheumatic diseases. *Curr Drug Targets* 2006; 7:91-102.

[13] Wang CR, Liu MF. Regulation of CCR5 expression and MIP-1α production in $CD4^+$T cells from patients with rheumatoid arthritis. *Clin Exp Immunol* 2003; 132:371-378.

[14] Volin MV, Woods JM, Amin MA, Connors MA, Harlow LA, Koch AE. Fractalkine: a novel angiogenic chemokine in rheumatoid arthritis. *Am J Pathol* 2001; 159:1521-1530.

[15] Wang CR, Liu MF, Huang YH, Chen HC: Upregulation of XCR1 expression in rheumatoid joint. *Rheumatology* 2004; 43:569-574.

[16] Wang CR. Involvement of natural killer T cells in C57BL/6 mouse model of collagen-induced arthritis. *J Microbiol Immunol Infect* 2003; 36:15-20.

[17] Meyer A, Coyle AJ, Proudfoot AEI, Wells TNC, Power CA. Cloning and characterization of a novel murine macrophage inflammatory protein-1 receptor. *J Biol Chem* 1996; 271:14445-14451.

[18] Huang H, Li F, Gordon JR, Xiang J. Synergistic enhancement of antitumor immunity with adoptively transferred tumor-specific $CD4^+$ and $CD8^+$ T cells and intratumoral lymphotactin transgene expression. *Cancer Res* 2002; 62:2043-2051.

[19] Hsieh JL, Wu CL, Lee CH, Shiau AL. Hepatitis B virus X protein sensitizes hepatocellular carcinoma cells to cytolysis induced by E1B-deleted adenovirus through the disruption of p53 function. *Clin Cancer Res* 2003; 9:338-345.

[20] Bristow CA, Shore P. Transcriptional regulation of the human MIP-1 promoter by RUNX1 and MOZ. *Nucleic Acids Res* 2003; 31:2735-2744.

[21] Wang CR, Chen SY, Wu CL, et al. Prophylactic adenovirus-mediated human kallistatin gene therapy suppresses rat arthritis by inhibiting angiogenesis and inflammation. *Arthritis Rheum* 2005; 52:1319-1324.

[22] Luther A, Cyster JG. Chemokines as regulators of T cell differentiation. *Nat Immunol* 2001; 2:102-107.

[23] Shiau AL, Lin PR, Chang MY, Wu CL. Retrovirus-mediated transfer of prothymosin gene inhibits tumor growth and prolongs survival in murine bladder cancer. *Gene Ther* 2001; 8:1609-1617.

[24] Kostrikis LG, Huang Y, Moore JP, et al. A chemokine receptor CCR2 allele delays HIV-1 disease progression and is associated with a CCR5 promoter mutation. *Nat Med* 1998; 4:350-353.

[25] McDermott DH, Zimmerman PA, Guignard F, Kleeberger CA, Leitman SF, Murphy PM. The Multicenter AIDS Cohort Studies: CCR5 promoter polymorphism and HIV-1 disease progression. *Lancet* 1998; 352:866-870.

[26] Samson M, Libert F, Doranz, BJ, et al. Resistance to HIV-1 infection in Caucasian individuals bearing mutant alleles of the CCR5 chemokine receptor gene. *Nature* 1996; 382:722-725.

[27] Cocchi F, DeVico AL, Garzino-Demo A, Arya SK, Gallo RC, Lusso P. Identification of RANTES, MIP-1 alpha, and MIP-1 beta as the major HIV-suppressive factors produced by CD8+ T cells. *Science* 1995; 270:1811-1815.

[28] Deighton CM, Walker DJ, Griffiths ID, Roberts DF. The contribution of HLA to rheumatoid arthritis. *Clin Genet* 1989; 36:178-182.

[29] Gomez-Reino JJ, Pablos JL, Carreira PE, et al. Association of rheumatoid arthritis with a functional chemokine receptor CCR5. *Arthritis Rheum* 1999; 42:989-992.

[30] Wang CR, Guo HR, Liu MF. RANTES promoter polymorphism as a genetic risk factor of rheumatoid arthritis in Chinese. *Clin Exp Rheumatol* 2005; 23:379-384.

[31] Liblau RS, Singer SM, McDevit HO. Th1 and Th2 $CD4^{+}T$ cells in pathogenesis of organ-specific autoimmune diseases. *Immunol Today* 1995; 16:34-38.

[32] Mauri C, Williams RO, Walmsley M. Relation between Th1 and Th2 cytokine patterns and the arthritogenic response in collagen-induced arthritis. *Eur J Immunol* 1996; 26:1511-1518.

[33] Quinones MP, Estrada CA, Kalkonde Y, et al. The complex role of the chemokine receptor CCR2 in collagen-induced arthritis: implications for therapeutic targeting of CCR2 in rheumatoid arthritis. *J Mol Med* 2005; 83:672-681.

[34] Lederman MM, Veazey RS, Offord R, et al. Prevention of vaginal SIV transmission in rhesus macaques through inhibition of CCR5. Science 2004; 306:485-487.

[35] Plater-Zyberk C, Hoogewerf J, Proudfoot A, Power C, Wells, TNC. Effect of a CC chemokine receptor antagonist on collagen induced arthritis in DBA/1 mice. *Immunol Lett* 1997; 57:117-120.

[36] Vierboom MPM, Zavodny PJ, Chou CC, et al. Inhibition of the development of collagen-induced arthritis in rhesus monkeys by a small molecular weight antagonist of CCR5. *Arthritis Rheum* 2005; 52:627-636.

[37] Okamoto H, Kamatani NA. CCR-5 antagonist inhibits the development of adjuvant arthritis in rats. *Rheumatology* 2006; 45:230-232.

[38] Bao L, Zhu Y, Zhu J, Lindgren JU. Decreased IgG production but increased MIP-1beta expression in collagen-induced arthritis in C-C chemokine receptor 5-deficient mice. *Cytokine* 2005; 31:64-71.

[39] Muller S, Dorner B, Korthauer U, et al. Cloning of ATAC, an activation-induced, T cell-derived, and chemokine-related molecule exclusively expressed in CD8+ T lymphocytes. *Eur J Immunol* 1995; 25:1744-1748.

[40] Blaschke S, Middel P, Dorner BG, et al. Expression activation-induced, T cell-derived, and chemokine-related cytokine/lymphotactin and its functional role in rheumatoid arthritis. *Arthritis Rheum* 2003; 48:1858-1872.

[41] Dorner BG, Scheffold A, Rolph MS, et al. MIP-1alpha, MIP-1beta, RANTES and ATAC/lymphotactin function together with IFN-gamma as type 1 cytokines. *Proc Natl Acad Sci USA* 2002; 99:6181-6186.

[42] Koch AE, Kunkle SL, Shah MR *et al.* Macrophage inflammatory protein-1 alpha, a novel chemotactic cytokine for macrophage in rheumatoid arthritis. *J Clin Invest* 1994; 93:921-928.

[43] Kasama T, Strieter RM, Lukacs NW, Lincoln PM, Burdick MD. Interleukin-10 expression and chemokine regulation during the evolution of murine type II collagen-induced arthritis. *J Clin Invest* 1995; 95:2868-2876.

[44] Shahrara S, Amin AM, Woods, JM, Haines GK, Koch AE. Chemokine receptor expression and in vivo signaling pathways in the joints of rats with adjuvant-induced arthritis. *Arthritis Rheum* 2003; 48:3568-3583.

[45] Kagari T, Doi H, Shimozato T. The importance of IL-1 and TNF , and no involvement of IL-6 in the development of monoclonal antibody-induced arthritis. *J Immunol* 2002; 169: 1459-1466.

[46] Evans CH, Ghivizzani SC, Robbins PD. Gene therapy for arthritis. What next? *Arthritis Rheum* 2006; 54:1714-1729.

[47] Evans CH, Ghivizzani SC, Robbins PD, et al. Gene transfer to human joints: progress toward a gene therapy of arthritis. *Proc Natl Acad Sci USA* 2005; 102:8698-8703.

[48] O'Dell RJ. Therapeutic strategies for rheuma-toid arthritis. *New Eng J Med* 2004; 350:2591-2602.

[49] Haringman JJ, Gerlag DM, Smeets TJ, et al. A randomized controlled trial with an anti-CCL2 (anti-monocyte chemotactic protein 1) monoclonal antibody in patients with rheumatoid arthritis. *Arthritis Rheum* 2006; 54:2387-2392.

[50] Haringman JJ, Kraan MC, Smeets TJ, Zwinderman KH, Tak PP. Chemokine blockade and chronic inflammatory disease: proof of concept in patients with rheumatoid arthritis. *Ann Rheum Dis* 2003; 62:715-721.

In: Chemokine Research Trends
Editor: Leah R. Grinwald, pp. 107-119

ISBN: 978-60021-588-9

Chapter 7

THE ROLE OF CHEMOKINE MCP-1 AND ITS RECEPTOR CCR2 IN NEUROPATHIC PAIN

Ji Zhang** *and Stefania Echeverry

Unité de Neurobiologie Cellulaire. Centre de Recherche Université Laval Robert-Giffard, 2601, Chemin de la Canardière. Québec, Québec, G1J 2G3. Canada

ABSTRACT

Neuropathic pain is the chronic pain state caused by pathological changes in the nervous system. It can occur secondarily to injury of the central nervous system but it occurs most commonly in association with injury to the peripheral nervous system. The pathogenesis of neuropathic pain involves neural-immune interactions that activate immune cells, glial cells and neurons. In this chapter, we summarize recent findings regarding the contribution of monocyte chemoattractant protein-1 (MCP-1), a CC family chemokine, in the etiology of this pathology. Constitutively present in the immune system, MCP-1 could be induced in damaged sensory neurons; its receptor, CCR2, is also present within the central nervous system. Under the powerful chemoattractant effect of MCP-1, monocytes/macrophages are recruited around the injured nerve and proinflammatory substances released by these immune cells increase the peripheral sensitization. In addition to its classical function in cell trafficking, MCP-1 could contribute to central sensitization by directly increase neuronal excitability or by modulation of spinal inhibitory GABAergic transmission. It is also suggested that MCP-1 could be an important molecule mediating neuron-glia interaction and trigger spinal glial activation. Binding of MCP-1 to CCR2 activates different signaling pathways upon the cellular environment and may define their physiopathological effect. These emerging observations provide new insights into the full spectrum of neuropathic pain and present continuing challenges for drug discovery.

ABBREVIATION

AIA: adjuvant induced-arthritis; ATF3: activating transcription factor-3, BBB: blood-brain barrier; CNS: central nervous system; DRG: dorsal root ganglion; EAE: experimental autoimmune encephalomyelitis; ERK: extracellular signal-regulated kinase, GFAP: glial fibrillary acidic protein; GPCR: G-protein coupled receptor; IL-1β: interleukin-1β; IL-6: interleukin-6, MCP-1: monocyte chemoattractant protein-1; MAPK/JNK: mitogen-activated protein kinase/c-Jun N-terminal kinase; PNS: peripheral nervous system; RANTES: regulated on activation normal T-cell expressed and secreted; PI3K: phosphoinositol-3-kinase; PTX: pertussis toxin, SDF-1: stromal cell derived factor-1, TNF-α: tumor necrosis factor-α,

INTRODUCTION

Neuropathic pain is a chronic sensory disorder associated with spontaneous abnormal sensation, as burning, stabbing, tingling, and/or abnormal stimulus-evoked pain, such as allodynia (pain sensation evoked by normally innocuous stimuli) and/or hyperalgesia (increased pain intensity evoked by normally painful stimuli). Neuropathic pain is caused by significant pathological changes in the nervous system. It can occur secondarily to injury of the central nervous system (CNS), but it occurs most frequently in association with injury to the peripheral nervous system (PNS). These injuries can be caused by trauma, tumors or disc herniation compressing peripheral nerves, metabolic disease as diabetes, viral disease as HIV, ischemic insults and toxins used for chemotherapy that stretches, compresses or inflames a nerve. The mechanisms underlying neuropathic pain are poorly understood and existing treatment is largely ineffective.

The pathogenesis of neuropathic pain states is very complex, involving structural, physiological and pharmacological changes throughout the neuroaxis (from the site of peripheral nerve injury to the spinal cord/brain). Beyond the unique priori about nerves or neurons involved, recent work has begun to illuminate some of the pathological process in the context of neuroinflammation. The interactions of immune system and pain system are realized at multiple levels, where many immune cells and immune-derived substances are implicated in the etiology of long-lasting neuropathic pain.

A family of protein chemoattractants called "chemokines" was discovered in the late 1980s. These small proteins (8-10 kDa) are classified, according to the number and spacing of the conserved N-terminal cysteines, into four classes (C, CC, CXC, CX3C). They act through specific or shared receptors that all belong to the superfamily of G-protein-coupled receptors (GPCR). The known chemokine system in humans comprises approximately 50 ligands and 20 GPCRs. The principal targets of chemokines are bone marrow-derived cells. Because motility is an essential part of their function, chemokines play a central role in leucocyte physiology by controlling basal and inflammatory trafficking [1,2]. A general view evolved in which leucocyte infiltration in disease could be explained by a cascade of mechanistic steps: a pathogenic stimulus induces chemokine secretion by cells in specific anatomic sites, which attracts circulating leucocytes that express appropriate receptors into those sites; the activated leucocytes would, in certain cases, then cause damage (disease). As many other type cells also express chemokine receptors, including endothelia, smooth muscle cells, stromal cells, neurons and epithelial cells [3], in addition to localizing cells of the immune system to particular compartments, chemokines may be involved in other aspects of tissue homeostasis. It is well known that chemokines play an important role in the pathogenesis of neuroinflammatory disease states such as multiple sclerosis and rheumatoid arthritis [4,5].

* E-mail address: Ji.Zhang@crulrg,.ulaval.ca Tel: 1-418-663-5747 Fax: 1-418-663-8756

The scope of involvement of chemokines and their receptors has recently broadened considerably to extend beyond leukocyte chemoattractants in more-classical neuroimmunological disorders, and now includes novel actions that might contribute to the pathogenesis of neurodegenerative diseases such as Alzheimer's disease and/or AIDS dementia [6].

Several lines of evidence demonstrate that chemokines also contribute directly to nociception. Chemokine receptors CXCR4, CX3CR1, CCR4 and CCR5 co-localized with substance P and TRPV1 vanilloid receptor in DRG neurons [7]. Application of some chemokines to cultured rat DRG neurons increases intracellular [Ca^{2+}] in subsets of neurons that also respond to bradykinin, ATP or capsaisin, and results in the release of substance P. Subcutaneous injection of chemokines interleukin 8 (IL-8) [8], stromal cell derived factor-1 (SDF-1), regulated on activation normal T-cell expressed and secreted (RANTES) [], and fractalkine [9] produce allodynia and hyperalgesia. Pretreatment with SDF-1, followed by opioid administration into the periacqueductal gray matter of the brain, resulted in an increased rat tail flick response to painful stimulus, which suggests that activation of SDF-1 receptor CXCR4 could enhance the perception of pain, probably by desensitization of opioid receptor [10] Also, an *in vitro* study by Bezzi et al [11] revealed a cascade in which SDF-1 could alter astrocytic glutamate metabolism leading to the accumulation of extracellular glutamate which results in a sustained excitation of nearby neurons.

Biology of MCP-1/CCR2

MCP-1 (also termed CCL2) is a member of the CC chemokine family. It was identified as a monocyte-specific chemoattractant [12] that was later shown to also attract memory T lymphocytes and NK [13,14]. Because of its target cell specificity, MCP-1 was postulated to play a pathogenetic role in a variety of diseases characterized by mononuclear cell infiltration, including atherosclerosis, rheumatoid arthritis and multiple sclerosis [15-17]. MCP-1 is structurally and genetically related to other chemokines, such as MCP-2, -3, -4, and –5, that share similar properties and activate the same receptor, CCR2. However, the recent study by Lu et al [2] suggests that chemokine redundancy, as defined by receptor binding *in vitro*, may not be relevant *in vivo*, where specificity is achieved by timing and expression levels. This is also supported by the fact that MCP-3 does not substitute for MCP-1 in MCP-1 knock-out mice because it is not expressed by the tissues that normally express MCP-1 [2]. That is, when MCP-1 is absent, the other MCPs are not up-regulated to compensate for its absence. In addition, mice engineered to be deficient for MCP-1 alone have unique phenotypes, most of which match those seen in CCR2-knock out mice [2,18,19],

Although it is absent from normal CNS, MCP-1 was found to be induced in facial nucleus neurons by facial nerve transaction [20], in sympathetic ganglion neurons following post-ganglionic axotomy [21] and in dorsal root ganglion (DRG) neurons by sciatic nerve injury [22,23]. Recent studies have suggested a role for MCP-1/CCR2 in nociception. Both intraplantar and intrathecal injections of this chemokine have been shown to induce mechanical allodynia [22,24]. Mutant mice lacking the MCP-1 receptor CCR2 ($CCR2^{-/-}$) did not develop tactile allodynia after nerve injury [24]. Moreover, a pilot study has shown that systemic injection of a CCR2 antagonist, INCB3344 (Incyte Corp.) prevented experimental neuropathic pain in mice [25]. Thus, it is not clear whether CCR2 on spinal microglia induced by peripheral nerve injury [24] or CCR2 on peripheral monocytes/macrophages, or the synergy of both, is (are) responsible for the development of neuropathic pain. The target cells and the mechanisms involved for the MCP-1 produced by DRG neurons remain to be elucidated.

MECHANISMS BY WHICH MCP-1/CCR2 CONTRIBUTE TO THE DEVELOPMENT OF NEUROPATHIC PAIN

MCP-1 contributes to both peripheral and central sensitization by interaction with different types of cells. The main mechanisms by which MCP-1 enhances hypersensitivity following peripheral nerve injury are illustrated in Figure 1.

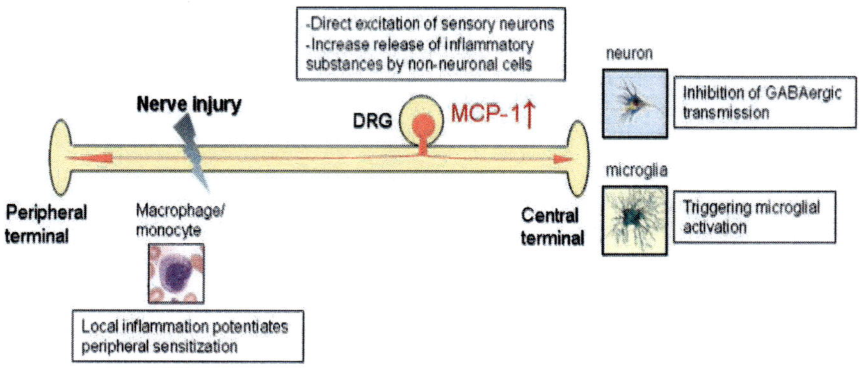

Figure 1. The main ways in which monocyte chemoattractant protein-1 (MCP-1) promotes hypersensitivity following peripheral nerve injury. Nerve injury induces MCP-1 expression in DRG sensory neurons, and the protein can be transported to their central and peripheral terminals. In the DRG, MCP-1 increases directly excitability of sensory neurons, and stimulates the release of TNF-α/IL-1β which excite indirectly sensory neurons. In the periphery, MCP-1 attracts monocytes and macrophages to the site of injury, and local inflammatory response potentiates peripheral sensitization. In the spinal cord dorsal horn, MCP-1 contributes to central sensitization by (1) triggering spinal microglial activation, and activated microglia modulate subsequently neuronal signal transmission in this neuropathic pain circumstances; (2) inhibition of spinal GABAergic transmission.

MCP-1 Induces Pain Hypersensitivity by Directly Exciting Primary Nociceptive Neurons

Numerous chemokines, including MCP-1, have been shown to elicit transient Ca^{2+} currents in dorsal root ganglion neurons [7]. The majority of DRG neurons (84.7%) that were sensitive to chemokines corresponded to nociceptors, because the same neurons also responded to the application of capsaicin, bradykinin or ATP [7], substances for which the effects typically are used as markers for nociceptive neurons [26-28]. MCP-1, absent in normal nervous system, could be induced in dorsal root ganglia sensory neurons following peripheral nerve injury [22,23]. Over half of MCP-1+ neurons were colocalized with TRPV1 (transient receptor potential vanilloid 1) [29]. A recent study addressed the role of MCP-1/CCR2 signaling in the context of chronic pain *in vivo* [30], they demonstrated that there is an up-regulation of CCR2 gene expression both in DRG neuronal and non-neuronal cells that is concurrent with the up-regulation of MCP-1 protein after a chronic compression injury. The application of MCP-1 to compressed DRG neurons exhibited potent excitatory effects via activation of a nonvoltage-dependent depolarizing current and inhibition of a voltage-dependent outward current [31]. These results suggest that *in situ* release of MCP-1 by damaged neurons can act as a neurotransmitter driving the hyperexcitability of nociceptive neurons in a cellular paracrine or autocrine manner. As CCR2 receptor was also present in most of the non neuronal cells around neurons in the injured DRG, it is possible that MCP-1 acted on non-neuronal cells to release factors such as TNF-α, or IL-1β that excite indirectly the DRG neurons.

Involvement of MCP-1/CCR2 in Macrophages Recruitment Following Injury to Peripheral Nerve

Following injury to the peripheral nervous system, circulating monocytes/macrophages are recruited to the damaged tissue, their numbers peaking at an early stage during the intense period of phagocytic activity. The invading macrophages secrete components of the complement cascade, coagulation factors, proteases, hydrolyses, interferons, tumor necrosis factor-α (TNF-α) and other cytokines to potentiate the process of proximal-distal axonal and Schwann cell degeneration also known as Wallerian degeneration, which is tightly correlated with the development of neuropathic pain [32]. Macrophages significantly increase the concentration of proinflammatory cytokines and many other proinflammatory substances [33]. It is also believed that local liberation of TNF-α at the site of nerve injury causes spontaneous electrophysiological activity in nociceptive and other sensory fibers [34]. Blocking the upregulation of TNF- α, interleukin-1β (Il-1β) and interleukin-6 (Il-6) or the recruitment of macrophages after nerve injury can interfere with the rate and magnitude of Wallerian degeneration, influencing the magnitude and the duration of the associated pain state [35,36].

The mechanisms by which monocytes/macrophages are recruited to the site of nerve injury are not entirely known. It is thought that adhesion molecules play a central role in promoting transendothelial migration. However, many recent data support the critical role of MCP-1 in macrophage recruitment to the injured peripheral nervous system. In the peripheral nervous system, increased concentrations of MCP-1 have been observed after nerve transection [37,38] and tellurium-induced demyelination [39]. MCP-1 has been shown to be necessary for full recruitment of monocytes during Wallerian degeneration. In mice with null deletion of CCR2 receptor [40] or treated with blocking antibodies [41], the concentration of macrophages in their nerves after injury is significantly reduced. Injection of recombinant MCP-1 into dorsal column of the spinal cord undergoing Wallerian degeneration triggers rapid macrophage activation [41]. The fact that CCR2 antagonist (INCB3344) treatment results in a dose dependent inhibition of macrophage influx in thioglycolate-induced peritonitis and its efficacy in mouse experimental autoimmune encephalomyelitis (EAE) model and in rat adjuvant induced-arthritis (AIA) model strongly support the role of MCP-1/CCR2 in macrophage/monocyte trafficking [25]. Besides its chemotactic activity, MCP-1 can stimulate production of cytokines by monocytes/macrophages [42] and induce histamine secretion from basophils [43].

MCP-1 Is Conveyed to the Central Terminals of the DRG Neurons and Act at the Spinal Cord Level

Our recent results implicate MCP-1 in spinal neuron-glia interactions following nerve injury. Indeed, although absent from the normal CNS, MCP-1 was induced in damaged activating-transcription factor-3 (ATF3)-positive DRG sensory neurons after peripheral nerve injury. The MCP-1 induced in DRG neurons was transported to afferent terminals in the spinal cord dorsal horn (as evidenced by co-localization with CGRP-positive terminals), whereas no microglial cells or neurons resident in the dorsal horn appeared to express MCP-1 [23]. Time course analysis demonstrated that MCP-1 induction in the ipsilateral side DRG sensory neurons happened very early, only one day after sciatic nerve injury. After peaking at day 7, MCP-1 levels began to decline rapidly and had returned to baseline by day 150 [23]. In addition, induced neuronal MCP-1 colocalized with synaptophsin, a marker of neurotransmitter secretory vesicles, in the soma, axonal processes and distal terminals. MCP-1 could be released upon neuronal depolarization in a calcium-dependent manner in both the soma and distal terminals [44]. Consistent with these results, Thacker et al. [45] reported that, following electrical stimulation of the dorsal roots at C-fiber strength, the level of MCP-1 in spinal superfusate increased 300% above the basal level in rats with peripheral nerve ligation (day 4 after surgery) [45]. The fact that peripheral nerve injury can induce spinal microglial/astrocyte activation has been demonstrated in several models of neuropathic pain [23,46-48]. Microglial activation identified by OX-42 immunolabeling started from day 3 after nerve injury, peaked at day 7-14, and declined afterwards, whereas astrocytes labeled by glial fibrillary acidic protein (GFAP) demonstrated hypertrophic morphology from day 7 and reached a high level at day 22 following nerve lesion, and remained for a long period [23]. As the induction of MCP-1 occurred shortly after nerve injury and preceded the surrounding microglial activation, and CCR2 can be detected on spinal microglia after peripheral nerve injury [24], we suggested that MCP-1 induced on damaged neurons by peripheral nerve injury may serve as a trigger for local microglial activation. Deletion of MCP-1 gene delayed microglial activation in mice from cortical injury [49] and strongly supported our assumption. Substances released by activated microglia might contribute to the initiation of neuropathic pain and also participate in inducing delayed astrocyte activation. There is also strong evidence that MCP-1 is involved in the recruitment of both monocytes/macrophages and activated lymphocytes into the central nervous system during a variety of inflammation, infection and traumatic conditions [50-52]. The phenomena have been attributed to the potential role of MCP-1 in altering expression of tight junction-associated proteins in brain microvascular endothelial cells, thereby increasing the blood-brain barrier (BBB) permeability [53,54]. Bone marrow stem cells have the ability to populate the entire central nervous system into fully differenciated parenchymal microglia [55,56]. The role of bone-marrow derived microglia in neuropathic pain needs to be elucidated in further experiments.

Although it has been previously recognized that CCR2 expression in the CNS is induced during the course of acute or chronic pathological conditions in astrocytes and microglial cells [24,57], recent studies also report constitutive expression of CCR2 in brain [58-60] and spinal cord neurons [61]. In investigating the function of neuronally expressed CCR2, Gosselin and his colleagues observed that MCP-1 potently and robustly inhibited GABA-induced neuronal currents [61]. They suggested a functional involvement of MCP-1/CCR2 in the modulation of spinal GABAergic transmission. As the main inhibitory neurotransmitter, GABA plays a major role in spinal cord physiology. One classical function exerted by GABA is its participation in the inhibitory modulation of somatosensory transmission, including nociception [62]. In particular, the blockade of spinal $GABA_A$ receptors has been shown to cause hyperalgesia and lead to tactile allodynia in the rat [63]. Thus, the potent inhibition of GABA-induced activation of $GABA_A$ receptors by MCP-1 might be one of the mechanisms by which the chemokine and its receptor, CCR2, facilitate nociceptive transmission at the spinal cord level.

Signaling Pathways

Studies of the signal transduction mechanisms of chemokine action have indicated that numerous chemokines trigger the activation signaling cascades classically associated with cell activation and mitogenesis, such as p42/44 mitogen-activated protein kinase/c-Jun N-terminal kinase (MAPK/JNK) and phosphoinositol-3-kinase (PI3K) pathways. The MAPKs are recognized as among the most versatile signaling kinases, which play an essential role in the regulation of crucial cellular functions in multiple cell types in response to a wide range of stimuli. CCR2 has been shown to couple to pertussis toxin (PTX) sensitive G-proteins, presumably $G_{\alpha i}$ [64], although it can transduce signals through other PTX-insensitive G proteins [65]. Intracellular calcium changes induced by MCP-1 are sensitive to PTX [66]. MCP-1 interacts with CCR2 to activate different MAPK cascades [67] that are important for the physiological function of MCP-1 and have been recently shown to be involved in intergrin activation and chemotaxis triggered by MCP-1 [68]. Extracellular signal-regulated kinases (ERK) activation by MCP-1 involves G_i, protein kinases C, PI3K, and Ras [69]. Sodhi and Biswas [70] showed that MCP-1-induced phosphorylation of $P42/44^{ERK}$ leads to the activation of murine peritoneal macrophages. P38 MAPK participation in MCP-1-stimulated migration of monocytes was demonstrated by using the P38 inhibitor, while the inhibition of $P42/44^{ERK1/2}$ MAPKs was ineffective [71]. On the contrary, the MCP-1-stimulated chemotaxis of endothelial cells was completely abolished by inhibitors of PI3K and $P42/44^{ERK1/2}$ MAPKs, but not by P38 MAPK inhibitors [72]. MCP-1 also triggers tyrosine phosphrylation and activation of JAK/STAT pathway in a PTX-independent manner [73]. This was confirmed by an *in vivo* study which demonstrated that CCR2 expression and tyrosine phosphorylation increases with peak inflammation in rat adjuvant-induced arthritis (AIA), a model for rheumatoid arthritis, and CCR2 tyrosine phosphorylation was associated with the JAK/STAT pathway at different stages of rat AIA, as well as macrophage and endothelial cell infiltration [74].

Parallel signal transduction pathways are activated by MCP-1. Depending on the cell type, these pathways differentially contribute to cell chemotactic and other activities. All these data may help in identification of possible molecular targets for attenuation of MCP-1 action for the treatment of various diseases, including neuropathic pain.

CONCLUSION

It is likely that the involvement of MCP-1 in nerve injury induced-neuropathic pain might be at multiple levels. MCP-1 can act on excitatory pathways by sensitizing directly DRG sensory neurons. It can also block inhibitory GABAergic responses to potentiate nociceptive transmission. MCP-1, a molecule secreted by injured neurons might, serve as a mediateur in neuron-glia interaction and trigger glial activation. The contribution of activated glial cells in neuropathic pain has been documented by increasing evidence. In the periphery, MCP-1 is necessary in monocyte/macrophage recruitment during Wallerian degeneration. MCP-1, together with other pro-inflammatory cytokines released by macrophages around the damage nerve, can sensitize peripheral nociceptors. Reports of exciting data regarding the contribution of chemokines are just emerging, but promise to be the focus of intense research activities for years to come. It offers new avenues for drug discovery.

ACKNOWLEDGEMENT

This research was supported by Canadian Institutes for Health Research (CIHR) grant MOP-77624 to JZ

REFERENCES

[1] Gerard, C. and B. J. Rollins. 2001. Chemokines and disease. *Nat.Immunol.* 2:108-115.

[2] Lu, B., B. J. Rutledge, L. Gu, J. Fiorillo, N. W. Lukacs, S. L. Kunkel, R. North, C. Gerard, and B. J. Rollins. 1998. Abnormalities in monocyte recruitment and cytokine expression in monocyte chemoattractant protein 1-deficient mice. *J.Exp.Med.* 187:601-608.

[3] Rollins, B. J. 1997. Chemokines. *Blood* 90:909-928.

[4] Bajetto, A., R. Bonavia, S. Barbero, T. Florio, and G. Schettini. 2001. Chemokines and their receptors in the central nervous system. *Front Neuroendocrinol.* 22:147-184.

[5] De Benedetti, F., P. Pignatti, S. Bernasconi, V. Gerloni, K. Matsushima, R. Caporali, C. M. Montecucco, S. Sozzani, F. Fantini, and A. Martini. 1999. Interleukin 8 and monocyte chemoattractant protein-1 in patients with juvenile rheumatoid arthritis. Relation to onset types, disease activity, and synovial fluid leukocytes. *J.Rheumatol.* 26:425-431.

[6] Bajetto, A., R. Bonavia, S. Barbero, and G. Schettini. 2002. Characterization of chemokines and their receptors in the central nervous system: physiopathological implications. *J.Neurochem.* 82:1311-1329.

[7] Oh, S. B., P. B. Tran, S. E. Gillard, R. W. Hurley, D. L. Hammond, and R. J. Miller. 2001. Chemokines and glycoprotein120 produce pain hypersensitivity by directly exciting primary nociceptive neurons. *J.Neurosci.* 21:5027-5035.

[8] Cunha, F. Q., B. B. Lorenzetti, S. Poole, and S. H. Ferreira. 1991. Interleukin-8 as a mediator of sympathetic pain. *Br.J.Pharmacol.* 104:765-767.

[9] Watkins, L. R. and S. F. Maier. 2002. Beyond neurons: evidence that immune and glial cells contribute to pathological pain states. *Physiol Rev.* 82:981-1011.

[10] Steele, A. D., I. Szabo, F. Bednar, and T. J. Rogers. 2002. Interactions between opioid and chemokine receptors: heterologous desensitization. *Cytokine Growth Factor Rev.* 13:209-222.

[11] Bezzi, P., M. Domercq, L. Brambilla, R. Galli, D. Schols, E. De Clercq, A. Vescovi, G. Bagetta, G. Kollias, J. Meldolesi, and A. Volterra. 2001. CXCR4-activated astrocyte glutamate release via TNFalpha: amplification by microglia triggers neurotoxicity. *Nat.Neurosci.* 4:702-710.

[12] Matsushima, K., C. G. Larsen, G. C. DuBois, and J. J. Oppenheim. 1989. Purification and characterization of a novel monocyte chemotactic and activating factor produced by a human myelomonocytic cell line. *J.Exp.Med.* 169:1485-1490.

[13] Loetscher, P., M. Seitz, M. Baggiolini, and B. Moser. 1996. Interleukin-2 regulates CC chemokine receptor expression and chemotactic responsiveness in T lymphocytes. *J.Exp.Med.* 184:569-577.

[14] Carr, M. W., S. J. Roth, E. Luther, S. S. Rose, and T. A. Springer. 1994. Monocyte chemoattractant protein 1 acts as a T-lymphocyte chemoattractant. *Proc.Natl.Acad.Sci.U.S.A* 91:3652-3656.

[15] Nelken, N. A., S. R. Coughlin, D. Gordon, and J. N. Wilcox. 1991. Monocyte chemoattractant protein-1 in human atheromatous plaques. *J.Clin.Invest* 88:1121-1127.

[16] Koch, A. E., S. L. Kunkel, L. A. Harlow, B. Johnson, H. L. Evanoff, G. K. Haines, M. D. Burdick, R. M. Pope, and R. M. Strieter. 1992. Enhanced production of monocyte chemoattractant protein-1 in rheumatoid arthritis. *J.Clin.Invest* 90:772-779.

[17] Ransohoff, R. M., T. A. Hamilton, M. Tani, M. H. Stoler, H. E. Shick, J. A. Major, M. L. Estes, D. M. Thomas, and V. K. Tuohy. 1993. Astrocyte expression of mRNA encoding cytokines IP-10 and JE/MCP-1 in experimental autoimmune encephalomyelitis. *FASEB J.* 7:592-600.

[18] Boring, L., J. Gosling, S. W. Chensue, S. L. Kunkel, R. V. Farese, Jr., H. E. Broxmeyer, and I. F. Charo. 1997. Impaired monocyte migration and reduced type 1 (Th1) cytokine responses in C-C chemokine receptor 2 knockout mice. *J.Clin.Invest* 100:2552-2561.

[19] Kuziel, W. A., S. J. Morgan, T. C. Dawson, S. Griffin, O. Smithies, K. Ley, and N. Maeda. 1997. Severe reduction in leukocyte adhesion and monocyte extravasation in mice deficient in CC chemokine receptor 2. *Proc.Natl.Acad.Sci.U.S.A* 94:12053-12058.

[20] Flugel, A., G. Hager, A. Horvat, C. Spitzer, G. M. Singer, M. B. Graeber, G. W. Kreutzberg, and F. W. Schwaiger. 2001. Neuronal MCP-1 expression in response to remote nerve injury. *J.Cereb.Blood Flow Metab* 21:69-76.

[21] Schreiber, R. C., K. Krivacic, B. Kirby, S. A. Vaccariello, T. Wei, R. M. Ransohoff, and R. E. Zigmond. 2001. Monocyte chemoattractant protein (MCP)-1 is rapidly expressed by sympathetic ganglion neurons following axonal injury. *Neuroreport* 12:601-606.

[22] Tanaka, T., M. Minami, T. Nakagawa, and M. Satoh. 2004. Enhanced production of monocyte chemoattractant protein-1 in the dorsal root ganglia in a rat model of neuropathic pain: possible involvement in the development of neuropathic pain. *Neurosci.Res.* 48:463-469.

[23] Zhang, J. and Y. De Koninck. 2006. Spatial and temporal relationship between monocyte chemoattractant protein-1 expression and spinal glial activation following peripheral nerve injury. *J.Neurochem.* 97:772-783.

[24] Abbadie, C., J. A. Lindia, A. M. Cumiskey, L. B. Peterson, J. S. Mudgett, E. K. Bayne, J. A. DeMartino, D. E. MacIntyre, and M. J. Forrest. 2003. Impaired neuropathic pain responses in mice lacking the chemokine receptor CCR2. *Proc.Natl.Acad.Sci.U.S.A* 100:7947-7952.

[25] Brodmerkel, C. M., R. Huber, M. Covington, S. Diamond, L. Hall, R. Collins, L. Leffet, K. Gallagher, P. Feldman, P. Collier, M. Stow, X. Gu, F. Baribaud, N. Shin, B. Thomas, T. Burn, G. Hollis, S. Yeleswaram, K. Solomon, S. Friedman, A. Wang, C. B. Xue, R. C. Newton, P. Scherle, and K. Vaddi. 2005. Discovery and pharmacological characterization of a novel rodent-active CCR2 antagonist, INCB3344. *J.Immunol.* 175:5370-5378.

[26] Cesare, P. and P. McNaughton. 1996. A novel heat-activated current in nociceptive neurons and its sensitization by bradykinin. *Proc.Natl.Acad.Sci.U.S.A* 93:15435-15439.

[27] Caterina, M. J., M. A. Schumacher, M. Tominaga, T. A. Rosen, J. D. Levine, and D. Julius. 1997. The capsaicin receptor: a heat-activated ion channel in the pain pathway. *Nature* 389:816-824.

[28] Gu, J. G. and A. B. Macdermott. 1997. Activation of ATP P2X receptors elicits glutamate release from sensory neuron synapses. *Nature* 389:749-753.

[29] Lamotte, R. H, Sun, J, White, F. A, Ya, N, and Ma, C. Chronic compression of the dorsal root ganglion potentiates neuronal responses to capsaicin via a chemokine-dependent mechanism. Society for Neuroscience 250.2/S26. 10-15-2006.

[30] White, F. A., J. Sun, S. M. Waters, C. Ma, D. Ren, M. Ripsch, J. Steflik, D. N. Cortright, R. H. Lamotte, and R. J. Miller. 2005. Excitatory monocyte chemoattractant protein-1 signaling is up-regulated in sensory neurons after chronic compression of the dorsal root ganglion. *Proc.Natl.Acad.Sci.U.S.A* 102:14092-14097.

[31] Sun, J. H., B. Yang, D. F. Donnelly, C. Ma, and R. H. Lamotte. 2006. MCP-1 enhances excitability of nociceptive neurons in chronically compressed dorsal root ganglia. *J.Neurophysiol.* 96:2189-2199.

[32] Myers, R. R., T. Yamamoto, T. L. Yaksh, and H. C. Powell. 1993. The role of focal nerve ischemia and Wallerian degeneration in peripheral nerve injury producing hyperesthesia. *Anesthesiology* 78:308-316.

[33] Leskovar, A., L. J. Moriarty, J. J. Turek, I. A. Schoenlein, and R. B. Borgens. 2000. The macrophage in acute neural injury: changes in cell numbers over time and levels of cytokine production in mammalian central and peripheral nervous systems. *J.Exp.Biol.* 203:1783-1795.

[34] Sorkin, L. S., W. H. Xiao, R. Wagner, and R. R. Myers. 1997. Tumour necrosis factor-alpha induces ectopic activity in nociceptive primary afferent fibres. *Neuroscience* 81:255-262.

[35] Campana, W. M., X. Li, V. I. Shubayev, M. Angert, K. Cai, and R. R. Myers. 2006. Erythropoietin reduces Schwann cell TNF-alpha, Wallerian degeneration and pain-related behaviors after peripheral nerve injury. *Eur.J.Neurosci.* 23:617-626.

[36] Romero-Sandoval, A. and J. C. Eisenach. 2006. Perineural clonidine reduces mechanical hypersensitivity and cytokine production in established nerve injury. *Anesthesiology* 104:351-355.

[37] Taskinen, H. S. and M. Roytta. 2000. Increased expression of chemokines (MCP-1, MIP-1alpha, RANTES) after peripheral nerve transection. *J.Peripher.Nerv.Syst.* 5:75-81.

[38] Subang, M. C. and P. M. Richardson. 2001. Influence of injury and cytokines on synthesis of monocyte chemoattractant protein-1 mRNA in peripheral nervous tissue. *Eur.J.Neurosci.* 13:521-528.

[39] Toews, A. D., C. Barrett, and P. Morell. 1998. Monocyte chemoattractant protein 1 is responsible for macrophage recruitment following injury to sciatic nerve. *J.Neurosci.Res.* 53:260-267.

[40] Siebert, H., A. Sachse, W. A. Kuziel, N. Maeda, and W. Bruck. 2000. The chemokine receptor CCR2 is involved in macrophage recruitment to the injured peripheral nervous system. *J.Neuroimmunol.* 110:177-185.

[41] Perrin, F. E., S. Lacroix, M. Aviles-Trigueros, and S. David. 2005. Involvement of monocyte chemoattractant protein-1, macrophage inflammatory protein-1alpha and interleukin-1beta in Wallerian degeneration. *Brain* 128:854-866.

[42] Reape, T. J. and P. H. Groot. 1999. Chemokines and atherosclerosis. *Atherosclerosis* 147:213-225.

[43] Uguccioni, M., C. R. Mackay, B. Ochensberger, P. Loetscher, S. Rhis, G. J. LaRosa, P. Rao, P. D. Ponath, M. Baggiolini, and C. A. Dahinden. 1997. High expression of the chemokine receptor CCR3 in human blood basophils. Role in activation by eotaxin, MCP-4, and other chemokines. *J.Clin.Invest* 100:1137-1143.

[44] Jung, H, Toth P.T, White, F. A, and Miller, R. J. The chemokine monocyte chemoattractant protein-1 (MCP-1/CCL2) functions as a neurotransmitte in dorsal root ganglia. Society for Neuroscience 250.4/T2. 10-15-2006.

[45] Ref Type: Abstract

[46] Thacker, M, Clark, A. C., and McMahon, S. B. The b chemokine CCL2 induces pain behaviours and spinal cord microglial activation in the adult Wistar rat. 2005 Abstract Viewer, Sydney, Australia: International Association for the Study of PAin Program No. 37-P14. 8-22-2005.

[47] Ref Type: Abstract

[48] 46. Colburn, R. W., A. J. Rickman, and J. A. DeLeo. 1999. The effect of site and type of nerve injury on spinal glial activation and neuropathic pain behavior. *Exp.Neurol.* 157:289-304.

[49] 47. Fu, K. Y., A. R. Light, G. K. Matsushima, and W. Maixner. 1999. Microglial reactions after subcutaneous formalin injection into the rat hind paw. *Brain Res.* 825:59-67.

[50] 48. Zhang, J., C. Hoffert, H. K. Vu, T. Groblewski, S. Ahmad, and D. O'Donnell. 2003. Induction of CB2 receptor expression in the rat spinal cord of neuropathic but not inflammatory chronic pain models. *Eur.J.Neurosci.* 17:2750-2754.

[51] 49. Muessel, M. J., R. M. Klein, A. M. Wilson, and N. E. Berman. 2002. Ablation of the chemokine monocyte chemoattractant protein-1 delays retrograde neuronal degeneration, attenuates microglial activation, and alters expression of cell death molecules. *Brain Res.Mol.Brain Res.* 103:12-27.

[52] 50. Kelder, W., J. C. McArthur, T. Nance-Sproson, D. McClernon, and D. E. Griffin. 1998. Beta-chemokines MCP-1 and RANTES are selectively increased in cerebrospinal fluid of patients with human immunodeficiency virus-associated dementia. *Ann.Neurol.* 44:831-835.

[53] Rancan, M., V. I. Otto, V. H. Hans, I. Gerlach, R. Jork, O. Trentz, T. Kossmann, and M. C. Morganti-Kossmann. 2001. Upregulation of ICAM-1 and MCP-1 but not of MIP-

2 and sensorimotor deficit in response to traumatic axonal injury in rats. *J.Neurosci.Res.* 63:438-446.
[54] Huang, D. R., J. Wang, P. Kivisakk, B. J. Rollins, and R. M. Ransohoff. 2001. Absence of monocyte chemoattractant protein 1 in mice leads to decreased local macrophage recruitment and antigen-specific T helper cell type 1 immune response in experimental autoimmune encephalomyelitis. *J.Exp.Med.* 193:713-726.
[55] Song, L. and J. S. Pachter. 2004. Monocyte chemoattractant protein-1 alters expression of tight junction-associated proteins in brain microvascular endothelial cells. *Microvasc.Res.* 67:78-89.
[56] Stamatovic, S. M., R. F. Keep, S. L. Kunkel, and A. V. Andjelkovic. 2003. Potential role of MCP-1 in endothelial cell tight junction 'opening': signaling via Rho and Rho kinase. *J.Cell Sci.* 116:4615-4628.
[57] Simard, A. R. and S. Rivest. 2004. Bone marrow stem cells have the ability to populate the entire central nervous system into fully differentiated parenchymal microglia. *FASEB J.* 18:998-1000.
[58] Simard, A. R., D. Soulet, G. Gowing, J. P. Julien, and S. Rivest. 2006. Bone marrow-derived microglia play a critical role in restricting senile plaque formation in Alzheimer's disease. *Neuron* 49:489-502.
[59] Fife, B. T., G. B. Huffnagle, W. A. Kuziel, and W. J. Karpus. 2000. CC chemokine receptor 2 is critical for induction of experimental autoimmune encephalomyelitis. *J.Exp.Med.* 192:899-905.
[60] Banisadr, G., F. Queraud-Lesaux, M. C. Boutterin, D. Pelaprat, B. Zalc, W. Rostene, F. Haour, and S. M. Parsadaniantz. 2002. Distribution, cellular localization and functional role of CCR2 chemokine receptors in adult rat brain. *J.Neurochem.* 81:257-269.
[61] Banisadr, G., R. D. Gosselin, P. Mechighel, W. Rostene, P. Kitabgi, and P. S. Melik. 2005. Constitutive neuronal expression of CCR2 chemokine receptor and its colocalization with neurotransmitters in normal rat brain: functional effect of MCP-1/CCL2 on calcium mobilization in primary cultured neurons. *J.Comp Neurol.* 492:178-192.
[62] Banisadr, G., R. D. Gosselin, P. Mechighel, P. Kitabgi, W. Rostene, and S. M. Parsadaniantz. 2005. Highly regionalized neuronal expression of monocyte chemoattractant protein-1 (MCP-1/CCL2) in rat brain: Evidence for its colocalization with neurotransmitters and neuropeptides. *J.Comp Neurol.* 489:275-292.
[63] Gosselin, R. D., C. Varela, G. Banisadr, P. Mechighel, W. Rostene, P. Kitabgi, and S. Melik-Parsadaniantz. 2005. Constitutive expression of CCR2 chemokine receptor and inhibition by MCP-1/CCL2 of GABA-induced currents in spinal cord neurones. *J.Neurochem.* 95:1023-1034.
[64] Malcangio, M. and N. G. Bowery. 1996. GABA and its receptors in the spinal cord. *Trends Pharmacol.Sci.* 17:457-462.
[65] Malan, T. P., H. P. Mata, and F. Porreca. 2002. Spinal GABA(A) and GABA(B) receptor pharmacology in a rat model of neuropathic pain. *Anesthesiology* 96:1161-1167.
[66] Baggiolini, M., B. Dewald, and B. Moser. 1997. Human chemokines: an update. *Annu.Rev.Immunol.* 15:675-705.
[67] Arai, H. and I. F. Charo. 1996. Differential regulation of G-protein-mediated signaling by chemokine receptors. *J.Biol.Chem.* 271:21814-21819.

[68] Aragay, A. M., M. Mellado, J. M. Frade, A. M. Martin, M. C. Jimenez-Sainz, A. Martinez, and F. Mayor, Jr. 1998. Monocyte chemoattractant protein-1-induced CCR2B receptor desensitization mediated by the G protein-coupled receptor kinase 2. *Proc.Natl.Acad.Sci.U.S.A* 95:2985-2990.

[69] Dubois, P. M., D. Palmer, M. L. Webb, J. A. Ledbetter, and R. A. Shapiro. 1996. Early signal transduction by the receptor to the chemokine monocyte chemotactic protein-1 in a murine T cell hybrid. *J.Immunol.* 156:1356-1361.

[70] Ashida, N., H. Arai, M. Yamasaki, and T. Kita. 2001. Distinct signaling pathways for MCP-1-dependent integrin activation and chemotaxis. *J.Biol.Chem.* 276:16555-16560.

[71] Jimenez-Sainz, M. C., B. Fast, F. Mayor, Jr., and A. M. Aragay. 2003. Signaling pathways for monocyte chemoattractant protein 1-mediated extracellular signal-regulated kinase activation. *Mol.Pharmacol.* 64:773-782.

[72] Sodhi, A. and S. K. Biswas. 2002. Monocyte chemoattractant protein-1-induced activation of p42/44 MAPK and c-Jun in murine peritoneal macrophages: a potential pathway for macrophage activation. *J.Interferon Cytokine Res.* 22:517-526.

[73] Ashida, N., H. Arai, M. Yamasaki, and T. Kita. 2001. Differential signaling for MCP-1-dependent integrin activation and chemotaxis. *Ann.N.Y.Acad.Sci.* 947:387-389.

[74] Arefieva, T. I., N. B. Kukhtina, O. A. Antonova, and T. L. Krasnikova. 2005. MCP-1-stimulated chemotaxis of monocytic and endothelial cells is dependent on activation of different signaling cascades. *Cytokine* 31:439-446.

[75] Mellado, M., J. M. Rodriguez-Frade, A. Aragay, G. del Real, A. M. Martin, A. J. Vila-Coro, A. Serrano, F. Mayor, Jr., and A. Martinez. 1998. The chemokine monocyte chemotactic protein 1 triggers Janus kinase 2 activation and tyrosine phosphorylation of the CCR2B receptor. *J.Immunol.* 161:805-813.

[76] Shahrara, S., M. A. Amin, J. M. Woods, G. K. Haines, and A. E. Koch. 2003. Chemokine receptor expression and in vivo signaling pathways in the joints of rats with adjuvant-induced arthritis. *Arthritis Rheum.* 48:3568-3583.

In: Chemokine Research Trends
Editor: Leah R. Grinwald, pp. 121-132
ISBN: 978-60021-588-9

Chapter 8

CCL2 AND CCL3 CHEMOKINES AND RANKL STIMULATE CHEMOTAXIS OF HUMAN MONOCYTES THROUGH DISTINCT INTRACELLULAR KINASE PATHWAYS

Fabrizio Montecucco, Maria Bertolotto, Giordano Bianchi, Franco Dallegri and Luciano Ottonello

Laboratory of Phagocyte Physiopathology, First Clinic of Internal Medicine, Department of Internal Medicine and Medical Specialties, University of Genoa Medical School, Genoa, Italy

ABSTRACT

Receptor activator of NF-kB ligand (RANKL) plays a crucial role in the molecular mechanisms of bone metabolism in bone-related diseases, such as osteoporosis and also in inflammatory conditions characterized by local bone remodelling including rheumatoid arthritis. In fact, RANKL is essential for differentiation of human monocytes into osteoclasts and also, it is a potent chemotactic factor for human monocytes with an efficacy similar to classic monocyte chemoattractants, such as CCL2 and CCL3. Nevertheless, the intracellular kinase pathways triggered by RANKL, CCL2 and CCL3 are still not well known and controversial. To clarify this issue, we investigated monocyte migration and kinase activation to the aforementioned ligands. Chemotaxis of human monocytes was performed in modified Boyden chamber with polycarbonate filter. Intracellular kinase activation was analyzed by western blot assay. Our results show that:

1. RANKL, CCL2 and CCL3 induce monocyte migration via a common pathway involving Src kinases, as suggested by the capacity of PP1 Analog (inhibitor of Src kinase) to inhibit monocyte migration to the three cytokines. On the contrary, although activated by each chemoattractant, ERK 1/2 activation is unnecessary for monocyte migration, as suggested by the lack of activity of PD 98059, the inhibitor of MEK 1/2 (a kinase activating ERK 1/2).

2. RANKL-induced monocyte locomotion is strictly related to PI3-K/Akt and p38 MAPK activity, but independent from JNK 1/2 activation. Indeed, LY 294002 (inhibitor of PI3-K), Akt inhibitor and SB 203580 (inhibitor of p38 MAPK) inhibited monocyte migration to RANKL, whereas SP600125 (inhibitor of JNK1/2) was ineffective. Accordingly, RANKL activated Akt and p38 MAPK, but not JNK 1/2..

3. CCL2 and CCL3 monocyte stimulation do not induce Akt, p38 MAPK and JNK 1/2 activation. Consistently, chemokine-driven monocyte migration is unaffected by LY 294002, Akt inhibitor, SB 203580, an SP600125.

These results provide evidence that RANKL- and CC chemokines-induced monocyte migration share a common pathway of signal transduction mediated by the activation of Src kinases. Nevertheless, RANKL delivers its locomotory signal via a transduction pathway downstream Src kinases, which is distinct from that of CCL2 and CCL3, PI3-K/Akt and p38 MAPK being selectively activated only by RANKL. The intracellular pathway(s) distal to Src and involved into CC chemokine-dependent migration remains to be identified. Thus, PI3-K/Akt and p38 MAPK may be proposed as possible targets for the pharmacologic inhibition of RANKL-mediated monocyte activation, without impairing monocyte locomotory response to classical chemokines generally involved in anti-infective responses.

INTRODUCTION

Monocytes are relatively inert circulating precursors of tissue macrophages endowed with poor phagocytic capacity [1]. In response to chemoattractant stimulation, monocytes transmigrate through endothelium using selectins and integrins and, after leaving the blood stream, they differentiate into macrophages, i.e. cells capable to phagocytose [1] and dendritic cells, i.e. professional antigen presenting cells [1]. These differentiated cells are among the major regulators of the immune response to foreign antigens, but, when their activity is disregulated, are implicated in the pathogenesis of various chronic inflammatory disorders. Classically, monocyte recruitment is thought to be driven by small cytokines endowed with chemotactic properties named chemokines [2]. It is generally assumed that the selective recruitment of inflammatory cells is regulated by the co-ordinated expression of the specific chemokine(s) at sites of inflammation and its/their corresponding receptor(s) on the cell surface. Thus, the pharmacologic interference with the chemokine/chemokine receptor system has been proposed as a rationale target for the therapy of various chronic inflammatory diseases [3].

Recruited monocytes are also capable of differentiating into osteoclasts, i.e. cells having a crucial role in regulating the bone remodeling process [4]. In this regard, it is well known that the development and the continuous remodeling of skeleton is regulated by inflammatory cytokines, growth factors, and hormones that contribute to the complex balance between the bone-forming and bone resorbing processes [4, 5]. In particular, recent findings support the role of receptor activator of nuclear factor (NF)κB ligand (RANKL) in the modulation of the synthesis of bone matrix by osteoblasts and the resorption of bone by osteoclasts [6, 7]. RANKL, also called tumor necrosis factor [TNF]-related activation-induced cytokine (TRANCE) or osteoprotegerin ligand (OPGL), is a cytokine and a new member of tumor-necrosis factor family [8] which interacts with RANK, a cell surface receptor, expressed by monocyte-derived osteoclast precursors, T cells, B cells and dendritic cells [9, 10]. Furthermore, recent evidence suggests that RANKL/RANK pathway provides crucial signal to attract circulating monocytes at the areas of severe bone destruction and, there, it promotes their differentiation into osteoclasts [4, 11-13]. Thus, at the present state of knowledge it seems that monocyte recruitment is regulated by not only classical inflammatory chemokines, such as CCL2 and CCL3, but also by non-classical chemoattractants such as RANKL, which is presently considered mainly involved in bone remodelling. Taking into account that osteoclasts have a relevant role in promoting localized bone remodeling during inflammatory diseases, including rheumatoid arthritis and periodontitis, rheumatoid arthritis being a major model of bone destructive inflammatory disease, it is suggestive that RANKL may represent a link between monocyte recruitment and bone destruction in such diseases. In this regard, the involvement of RANKL in the physiopathology of bone resorption during arthritis is also strongly supported by two other independent findings: 1. the treatment of osteoprotegerin (OPG), the decoy receptor of RANKL, prevents bone erosion in a model of adjuvant arthritis [14]; 2. the absence of osteoclast maturation from monocytic precursors in RANKL and RANK null mice results in a severe osteopetrosis [15, 16]. For these reasons, the blockade of RANKL might be a therapeutic option to control bone destruction in inflammatory diseases such as rheumatoid arthritis and periodontitis, characterized by a local increase of osteoclast activity and infiltration of monocytes and /or T cells [14]. However, although RANKL is essential for osteoclastogenesis, the binding with its receptor RANK triggers intricate and still unclear intracellular signaling cascades. On the basis of these premises, we investigated RANKL-mediated monocyte migration with the related intracellular kinases possibly involved, particularly Src kinases, PI3-K/Akt, p38 MAPK, ERK1/2, JNK 1/2, in comparison with the cellular responses induced by the classical chemokines CCL2 and CCL3.

Material and Methods

Purification of Monocytes

Human monocytes were purified, as previously described [17], from buffy coats of healthy volunteers after informed consent. After centrifugation on Ficoll-Hypaque (Seromed, Berlin, Germany) density gradient, mononuclear cells were collected from the interface and washed with PBS (Irvine Scientific, Santa Ana, CA). Then monocytes were purified from the upper interface of a hypotonic Percoll density gradient (1.129 g/ml) (Sigma-Aldrich S.r.l., Milano, Italy). Purified monocytes were resuspended in RPMI 1640 with 25 mM Hepes (Irvine Scientific) containing 1% (v/v) heat-inactivated fetal calf serum and 500 ng/ml polymyxin B. Viability of monocytes was more than 98% as determined by ethidium bromide-fluoresceine diacetate assay [18] and purity was at least 90% as assessed by flow cytometric analysis (CD14 staining) [19] and nonspecific esterase staining.

Monocyte Migration Assay

After the purification, monocytes were washed in PBS and then resuspended in chemotaxis medium (RPMI 1640 containing 25 mM Hepes, added with 1% bovine serum albumin (Sigma-Aldrich S.r.l.). Monocytes were then incubated for 1 hour at 37° C with 5% CO_2 in presence or in absence of the following kinase inhibitors: 4-Amino-1-teet-butyl-3-(1'-naphtyl)pyrazolo [3,4-α]pyrimidine (PP1 analog, Src kinase inhibitor, 20 μM, Calbiochem, San Diego, CA), LY294002 (PI3K inhibitor, 50 μM, Sigma-Aldrich S.r.l.), 1L-6-hydroxymethyl-*chiro*-inositol 2[(R)-2-*O*-methyl-3*O*-octadecylcaromate (Akt inhibitor, 40 μM, Calbiochem), SB203580 (p38 mitogen activated protein kinase (MAPK) inhibitor, 1 μM, Biomol Research Laboratories, Inc., Plymouth Meeting, PA), SP600125 (JNK inhibitor, 20 μM, Calbiochem) and PD98059 (MEK inhibitor, at 50 μM, Biomol Research Laboratories). After the incubation with kinase inhibitors, monocytes were washed with PBS and resuspended with in medium to test their locomotory response. Monocyte migration was assessed in a 48-well micro chemotaxis chamber (Neuro Probe Inc., Gaithersburg, MD) using a 5 μm pore size, 5-μm-thick polycarbonate polyvinylpyrrolidone-free filter. The lower wells of chemotaxis chamber were filled with 1 pM RANKL (PeproTech EC, London, UK), 10 nM CCL2, 10 nM CCL3 (both from PeproTech) or control medium and the upper well with 50 μL of cell suspension (1 x 10^6/ml). After incubation (60 min, 37°C), the filters were stained with Diff-Quick (Baxter, Rome, Italy). Each condition was performed in duplicate. The cells of 5 random oil-immersion fields were counted and the chemotaxis index was calculated from the number of cells migrated to the test samples divided by the number of cells migrated to the control.

Western Blot Analysis

Western Blot analysis was carried out as previously described [20]. After purification, monocytes (1 x 10^7 cells/ml) were incubated in air at 37° C in absence or in presence of 10 nM CCL2 for 3 minutes, 10 nM CCL3 for 5 minutes, 100 pM RANKL for 60 minutes and 5 ng/ml TNF-α (ICN Biomedicals Technical Information, Aurora, Ohio) for 10 minutes [11]. The reaction was stopped on ice and the cells were centrifugated at 4° C. After removing the supernatants, the pellets were lysed in 400 μl of NP40 buffer consisting of 20 mM Tris-HCl, pH 7.5, 150 mM NaCl, 10 mM NaF, 1% NP40, 10 μg/ml glycerol, 10 μg/ml aprotinin, 10 μg/ml leupeptin, 1 mM PMSF, 0.5 mM Na_3VO_4. The samples were then spun at 13,000 x g for 10 min at 4° C. Protein content was determined with BCA protein assay (Pierce, Rockford, IL) using bovine serum albumin as standard. The cytoplasmic extracts were boiled in loading buffer (62.5 mM Tris-HCl ph 6,8, 0,75% SDS, 3.75% 2-mercaptoethanol, 8.75% glycerol and 0.025% bromophenol blue) and then resolved by SDS-polyacrilamide electrophoresis (5% stacking gel and 15% running gel). After the electrophoresis, the proteins (40 μg each sample) were transferred on nitrocellulose membrane at 4° C for 45 min at 240 mA. Then, nitrocellulose membranes were stained with red Ponceau S to verify equal amounts of protein transferred for each sample. The blots were blocked with 5% non-fat dry milk and washed in Tris-buffered saline/Tween 20 (TBS-T, containing 10 mM tris-base ph 7.4, 154 mM NaCl and 0.05% tween 20). The membranes were stained with 1 μg/ml anti-human phosphorylated Akt mAb (Santa Cruz Biotechnology, Santa Cruz, CA), 0.5 μg/ml polyclonal anti-human phosphorylated p38 MAPK mAb (R&D System, Minneapolis, MN), 1 μg/ml monoclonal anti-human phosphorylated JNK 1/2 mAb (Santa Cruz Biotechnology) or 1 μg/ml monoclonal anti-human phosphorylated ERK 1/2 mAb (R&D System) (overnight incubation at 4° C). After washing with TBS-T, the membranes were incubated with the appropriate horseradish peroxidase-linked anti-IgG Ab (GE Healthcare, Chalfont St. Giles, UK). The membranes were washed again in the same conditions, and antibody complexes were visualized by the enhanced chemiluminescence (ECL) detection system (Amersham, Little Chalfont, U.K.) according to the manufacturer's instructions.

Densitometric and Statistical Analysis

Images of blots were analyzed by scanning densitometry quantified with the image analyzer system (Syngene, Frederick, MD). Data were expressed as mean ± SEM. One-way ANOVA with Bonferroni's post test was performed using GraphPad InStat version 3.05 for Windows XP, GraphPad Software (San Diego, CA). Differences between treatment groups were judged to be statistically significant at $p < 0.05$.

RESULTS

Human Monocytes Migrate in Response to RANKL and CCL2 and CCL3 Chemokines Through Src Kinases Activation

In order to investigate the role of various intracellular kinase pathways and on the basis of these previous experimental evidence [8], we tested monocyte migration in response to RANKL, CCL2 and CCL3, respectively at concentrations of 1 pM, 10 nM and 10 nM. We firstly examined the involvement of Src kinases family, known to mediate the activation of various downstream kinases involved in the regulation of cell migration. As shown in Fig 1, the pre-treatment with Src kinase inhibitor PP1 analog (20 μM) inhibited RANKL- , CCL2- and CCL3-induced monocyte migration, suggesting a crucial role for Src in the locomotory response of human monocyte to these different stimuli.

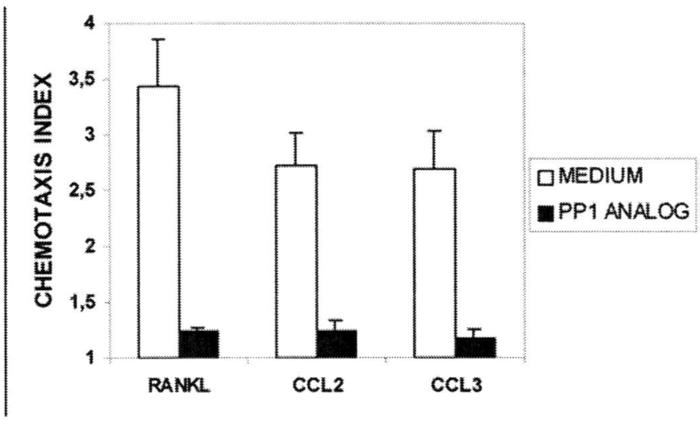

Figure 1. Effects of PP1 Analog (Src kinase selective inhibitor) on monocyte migration to chemoattractants. Monocytes pre-incubated with medium (open bars), or 20 μM PP1 Analog were tested for their ability to migrate to RANKL, CCL2 and CCL3. Data are expresses as chemotaxis index (mean ± 1 SEM, n = 6). Monocyte migration in absence vs. in presence of PP1 Analog in response to RANKL, CCL2 and CCL3: $p < 0.01$..

Role of PI3K/Akt on Monocyte Migration in Response to RANKL or CCL2 and CCL3 Chemokines

We next examined the involvement of PI3K and Akt in monocyte chemotaxis. As shown in Fig 2A, the pre-treatment with LY294002 (PI3K inhibitor), and Akt I (Akt inhibitor) inhibited monocyte migration in response to RANKL, whereas the two inhibitors were not able to impair the CCL2- and CCL3-induced monocyte chemotaxis. In agreement with these findings, western blot analysis shows that RANKL, but not CCL2 or CCL3, induces Akt activation in comparison with control medium (Fig 2B). Therefore, these data show that RANKL-induced migration of human monocytes is selectively dependent on PI3K-Akt activity, being these kinases not involved in monocyte migration to CCL2 and CCL3.

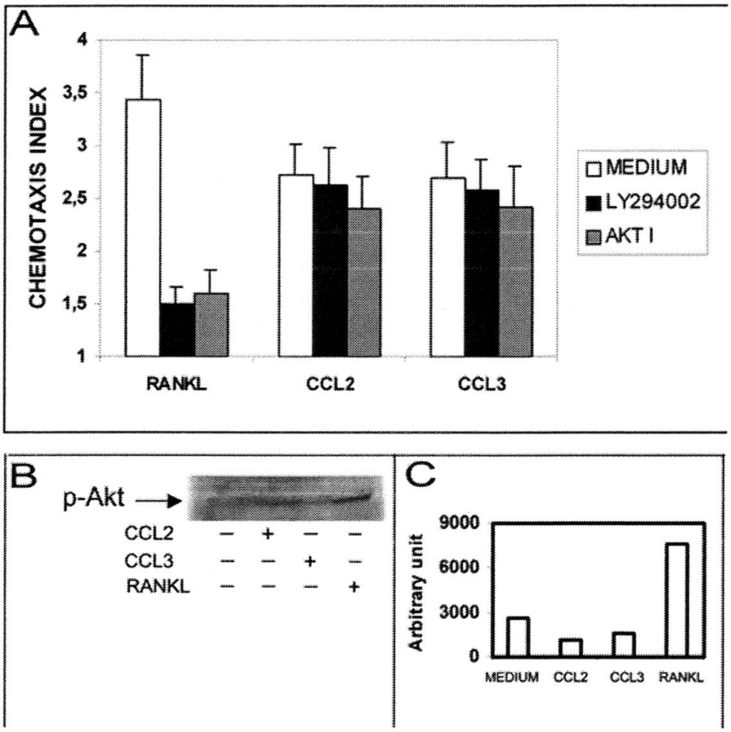

Figure 2. A, Effects of Akt inhibitor and LY294002 on monocyte migration to chemoattractants. Monocytes pre-incubated with medium (open bars), LY294002 (black bars) or Akt inhibitor (AKT I, (crossed bars) were tested for their ability to migrate to RANKL, CCL2 and CCL3. Data are expressed as chemotaxis index (mean ± 1 SEM, n = 6). Monocyte migration in absence vs. in presence of LY294002 or Akt inhibitor in response to RANKL: $p < 0.01$. Monocyte migration in absence vs. in presence of LY294002 or Akt inhibitor in response to CCL2 or CCL3 : NS. B, Western blot analysis of Akt phosphorylation. Monocytes were treated with control medium, CCL2 (10 nM, 5 min), CCL3 (10 nM, 3 min) or RANKL (1 pM, 60 min). Results represent one of three experiments that yielded similar results. C, Densitometric analysis of Western blot reported in panel B.

Role of MAPKs on Monocyte Migration in Response to RANKL or CCL2 and CCL3 Chemokines

We then studied the role and the activity of various MAPKs, i.e. p38 MAPK, ERK 1/2 and JNK 1/2 in monocytes exposed to RANKL and the CC chemokines. As shown in Fig. 3A RANKL, but not CCL2 or CCL3, induced p38 MAPK activation in comparison with control medium. On the contrary, the three ligands were able to activate ERK 1/2. Finally, RANKL, CCL2 and CCL3 were incapable of activating JNK 1/2, in comparison with the well detectable signal after stimulation with TNF-α, a well known activator of JNK 1/2 in monocytes [11]. To study the effective involvement of activated kinases in monocyte migration, a series of experiments with specific kinase inhibitors were performed.

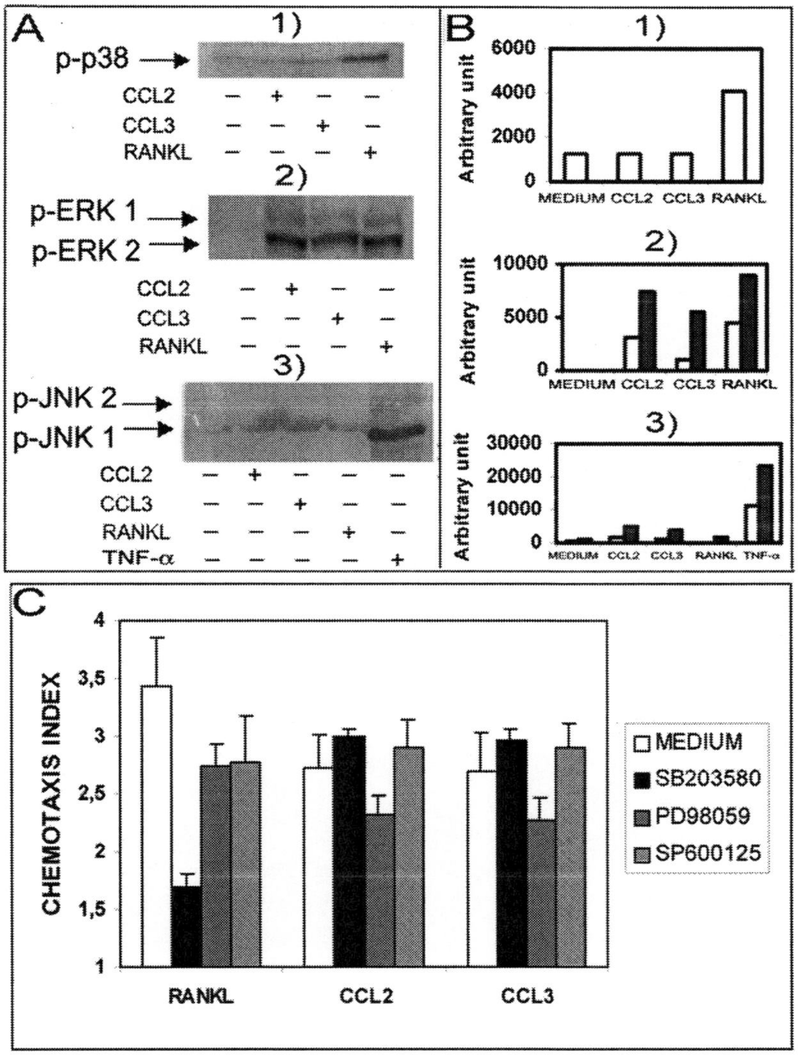

Figure 3. A, Western blot analysis of MAPK activation. Monocytes were treated with control medium, CCL2 (10 nM, 5 min), CCL3 (10 nM, 3 min), RANKL (1 pM, 60 min) or TNF-α (5 ng/ml, 10 min, used as positive control in JNK 1/2 analysis). Results represent one of three experiments that yielded similar results. B, Densitometric analysis of Western blot reported in panel. C, Effects of SB203580 (p38 MAPK inhibitor), PD98059 (MEK 1/2 selective inhibitor, a kinase activating ERK 1/2) or SP600125 (JNK 1/2 selective inhibitor) monocyte migration to chemoattractants. Monocytes pre-incubated with medium (open bars), SB 203580 (black bars), PD98059 (grey bars) or SP600125 (crossed bars) were tested for their ability to migrate to RANKL, CCL2 and CCL3. Data are expressed as chemotaxis index (mean ± 1 SEM, n = 6). Monocyte migration in absence vs. in presence of SB203580 in response to RANKL: $p < 0.01$. Other comparisons: NS.

As shown in Fig.3B, cell pre-treatment with SB203580 (a selective p38 MAPK inhibitor) significantly inhibited monocyte migration to RANKL, but not to CCL2 and CCL3. On the other hand, neither PD98059 (a selective inhibitor of MEK, a kinase activating ERK1/2) nor SP600125 (JNK 1/2 inhibitor) impaired RANKL-, CCL2- and CCL3- induced monocyte chemotaxis (Fig. 3B). This suggests that p38 MAPK activity is selectively necessary only for the RANKL-induced migration of human monocytes. Conversely, although activated by RANKL, CCL2 and CCL3 stimulation, ERK 1/2 is not involved on monocyte migration.

CONCLUSION

The pathophysiological model of bone erosion in chronic arthritis involving monocytes attributes a central role to RANKL and it is based on three main steps. First, monocytes, among other osteoclast precursor cells, enter the synovial membrane from the blood stream chemotactically recruited by RANKL [21], which is expressed in the rheumatoid arthritis synovial tissue (mainly in osteoblasts and stromal cells) [22]. In the second step the engagement of RANK by RANKL initiates the signaling cascade ultimately driving monocytes and macrophages to differentiate further into osteoclasts [23]. This stage is also supported by in situ production of macrophage-colony-stimulating factor, which in addition to RANKL [23] or TNF-α [11] is regarded as an essential molecule for osteoclasts differentiation. Finally, the third step is the bone resorption by mature and activated osteoclasts [9]. In this scenario, the interference with RANKL activity appears a possible target for an early and effective blockade of pathologic osteoclastogenesis. In this regard, a highly specific, a fully human antibody against RANKL has been recently produced capable of reducing the bone turnover by improving the bone density [24].

Our data confirm and extend previous experimental evidence [13, 21] showing that RANKL is a potent chemotactic factor for human monocytes, also when compared to the classical monocyte chemoattractants CCL2 and CCL3. The present finding confirms that Src kinase and PI3K inhibition blocks monocyte migration to RANKL [21]. Most importantly, we provide the first evidence that monocyte stimulation by RANKL results in Akt, p38 MAPK, ERK1/2, but not JNK 1/2 activation. Furthermore, the use of selective kinase inhibitors demonstrates that Akt and p38 MAPK are critical regulators of RANKL-dependent monocyte migration, whereas ERK 1/2 activity is irrelevant. On the contrary, PI3-K/Akt, p38 MAPK, ERK 1/2 and JNK 1/2 have no role in CCL2 or CCL3-induced chemotaxis, although Src kinase results involved. Reports regarding the secondary messengers involved in the signal transduction pathways mediated by CCL2 and CCL3 chemokines are quite controversial. In agreement with our findings, Arefieva and co-workers have previously show that Src kinases are essential for CCL2-mediated migration in human monocyte [25]. As far as PI3K involvement is concerned, although a study performed with murine macrophages [26] suggests a crucial role for this kinase in cell migration to CCL3, two independent works show that CCL3 mediates human monocyte chemotaxis without involving PI3K activation [27, 25]. Furthermore, Gerzten and co-workers do not observed PI3Kgamma activation in human monocytes, the isoform involved in leukocyte chemotaxis [28]. Previous data regarding MAPKs and CCL2- and CCL3-induced monocyte chemotaxis do not offer an univocal picture. Indeed, two reports [21, 25] [27, 25] show that ERK 1/2 activity is not required for monocyte migration toward CCL2 and CCL3. On the other hand, data from

Riboldi and co-workers indirectly suggest an involvement of ERK1/2 in monocyte migration to CCL2 [29]. Finally, Yen and co-workers have found two distinct pathways mediated by CCL2, one involving ERK1/2 activation, the other dependent on PI3K activation but independent from ERK1/2 activity [30]. As far as p38 MAPK and human monocytes is concerned, the only data available show a slight, although significant, inhibition of migration to CCL2 induced by p38 MAPK inhibition [25]. No data are reported about the role of JNK 1/2 in monocyte chemotaxis induced by these two chemokines, even if it was previously reported that CCL2 stimulates JNK 1 activation in a monoblastic cell line [31].

In summary, our results provide evidence that, although the activation of Src kinases is necessary for both RANKL- and CC chemokines-induced monocyte migration, PI3-K/Akt and p38 MAPK are selective targets to modulate only RANKL chemotactic property for human monocytes. Furthermore, the intracellular pathway(s) distal to Src governing monocyte migration to CCL2 and CCL3 remain to be identified. In this regard, possible candidates might be GTP binding proteins rho, rac and cdc42, that are known to be activated in CCL2- and CCL3-induced monocyte actin polarization [32-36]. Thus, our data shed new light on the selective targets linking bone metabolism and immunity for pharmacologic treatments to improve many diseases associated to bone loss. In particular, PI3-K/Akt and p38 MAPK may be proposed as possible targets for the pharmacologic inhibition of RANKL-mediated monocyte activation, without impairing monocyte locomotory response to classical chemokines generally involved in anti-infective responses.

REFERENCES

[1] Gordon, S. Development and distribution of mononuclear phagocytes. In: Gallin J, Snyderman R editors. *Inflammation: principles and clinical correlates, 3rd ed.* Philadelphia: Lippincott Williams & Wilkins; 1999; 35-48.

[2] Zlotnik, A. & Yoshie, O. (2000). Chemokines: a new classification system and their role in immunity. *Immunity,* **12**, 121–127.

[3] Johnson, Z.; Power, C.A.; Weiss, C.; Rintelen, F.; Ji, H.; Ruckle, T.; Camps, M.; Wells, T.N.C.; Schwarz, M.K.; Proudfoot, A.E.I. & Rommel, C. (2004). Chemokine inhibition: why, when, where, which and how? *Biochem. Soc. Trans.,* **32**, 366-377.

[4] Wada, T.; Nakashima, T.; Hiroshi, N. & Penninger, J. M. (2006). RANKL-RANK signaling in osteoclastogenesis and bone disease. *Trends Mol. Med.,* **12**, 17-25.

[5] Henriksen, K.; Karsdal, M.; Delaissè, J-M. & Engsig, M. T. (2003). RANKL and vascular endothelial growth factor (VEGF) induce osteocleast chemotaxis through an ERK 1/2-dependent mechanism. *J. Biol. Chem.,* **278**, 48745-48753.

[6] Boyle, W. J.; Simonet, W. S. & Lacey, D. L. (2003). Osteoclast differentiation and activation. *Nature,* **423**, 337-342.

[7] Karsenty, G. & Wagner, E. F. (2002). Reaching a genetic and molecular understanding of skeletal development. *Dev. Cell, 2,* **389**-406.

[8] Lacey, D. L.; Timms, E.; Tan, H.; Kelley, M. J.; Dunstan, C. R., Burgess, T.; Elliot, R.; Colombero, A.; Elliott, G.; Scully, S.; Hsu, H.; Sullivan, J.; Hawkins, N.; Davy, E.; Capparelli, C.; Eli, A.; Qian, Y. X.; Kaufman, S.; Sarosi, I.; Shalhoub, V.; Semaldi, G.; Guo, J.; Delaney, J. & Boyle, W. J. (1998). Osteoprotegerin ligand is a cytokine that regulates osteoclast differentiation and activation. *Cell,* **93**, 165-176.

[9] Hsu, H.; Lacey, D. L.; Dunstan, C. R.; Solovyev, I.; Colombero, A.; Timms, E.; , H. L.; Elliott, G; Kelley, M. J.; Sarosi, I.; Wang, L.; Xia, X. Z; Elliott, R.; Chiu, L.; Blank, T.; Scully, S.; Capparelli, C.; Morony, S.; Shimamoto, G.; Bass, M. B. & Boyle, W. J. (1999). Tumor necrosis factor receptor family member RANK mediates osteoclast differentiation and activation induced by osteoprotegerin ligand. *Proc. Natl. Acad. Sci. USA,* **96**, 3540-3545.

[10] Anderson, D. M.; Maraskovsky, E.; Billingsley, W. L.; Dougall, W. C., Tometsko, M. E.; Roux, E. R.; Teepe, M. C.; DuBose, R. F.; Cosman, D. & Galibert, L. (1997). A homologue of the TNF receptor and its ligand enhance T-cell growth and dendritic-cell function. *Nature,* **390**, 175-179.

[11] Kwak, H. B.; Jin, H-M.; Ha, H.; Kang, M-J.; Lee, B. S.; Kim, H-H. & Lee, Z. H. (2005). Tumor necrosis factor-α induces differentiation of human peripheral blood mononuclear cells into osteoclasts through the induction of $p21^{(WAF1/Cip1)}$. *Biochem Biophys Res Commun,* **330**, 1080-1086.

[12] Toh, K.; Kukita, T.; Wu, Z.; Kukita, A.; Sandra, F.; Tang, Q. Y.; Nomiyama, H. & Iijima, T. (2004). Possible involvement of MIP-1α in the recruitment of osteoclast progenitors to distal tibia in rats with adjuvant-induced arthritis. *Lab. Invest.,* **84**, 1092-1102.

[13] Breuil, V.; Schmid-Antomarchi, H.; Schmid-Alliana, A.; Rezzonizo, R.; Euller-Ziegler, L. & Rossi, B. (2003). The receptor activator of nuclear factor (NF)kB ligand (RANKL) is a new chemotactic factor for human monocytes. *FASEB J.,* **17** ,1751-1753.

[14] Kong, Y. Y.; Feige, U.; Sarosi, I.; Bolon B.; Tafuri, A.; Morony, S.; Capparelli, C.; Li, J.; Elliott, R.; McCabe, S.; Wong, T.; Campagnolo, G.; Moran, E.; Boqoch, E. R.; Van, G.; Nquyen, L. T.; Ohashi, P. S.; Lacey, D. L.; Fish, E.; Boyle, W. J. & Penninger, J. M. (1999). Activated T cells regulate bone loss and joint destruction in arthritis through osteoprotegerin ligand. *Nature,* **402**, 304-309.

[15] Takami, M.; Woo, J. T. & Nagai, K. (1999). Osteoblastic cells induce fusion and activation of osteoclasts through a mechanism independent of macrophage-colony-stimulating factor production. *Cell. Tissue Res.,* **298**, 327-334.

[16] Dougall, W. C.; Glaccum, M.; Charrier, K.; Rohrbach, K.; Brasel, K.; De Smedt, T.; Daro, E.; Smith, J.; Tometsko, M. E.; Maliszewski, C. R.; Armstrong, A.; Shen, V.; Bain, S.; Cosman, D.; Anderson, D.; Morrissey, P. J.; Peshon, J. J. & Schuh, J. (1999). RANK is essential fro osteoclast and lymph node development. *Genes Dev.,* **13**, 2412-2424.

[17] Ottonello, L.; Bertolotto, M.; Montecucco F.; Dapino, P. & Dallegri, F. (2005). Dexamethasone-induced apoptosis of human monocytes exposed to immune complexes. Intervention of CD95- and Xiap-dependent pathways. *Int. J. Immumopathol. Pharmacol.,* **18**, 403-415.

[18] Ottonello, L.; Frumento, G.; Arduino, N.; Bertolotto, M.; Dapino, P.; Mancini, M. & Dallegri, F. (2002). Differential regulation of spontaneous ad immune complexes-induced neutrophil apoptosis by proinflammatory cytokines. Role of oxidants, Bax and caspase-3. *J. Leukoc. Biol.,* **72**, 125-132.

[19] Mu, J. L.; Lanzarova, P.; Gaudernack, G.; Saeboe-Larssen, S. & Kvalheim, G. (2004). Development of a clinical grade procedure for generation of mRNA transfected

dendritic cells from purified frozen CD34+ blood progenitor cells. *Int. J. Immunopathol. Pharmacol.*, **17**, 255-263.

[20] Ottonello, L.; Montecucco, F.; Bertolotto, M.; Arduino, N.; Mancini, M.; Corcione, A.; Pistoia, V. & Dallegri, F. (2005). CCL3 (MIP-1alpha) induces in vitro migration of GM-CSF-primed human neutrophils via CCR5-dependent activation of ERK 1/2. *Cell. Signal.*, **17**, 355-363.

[21] Mosheimer, B. A.; Kaneider, N. C.; Feistritzer, C.; Sturn, D. H. & Wiedermann, C. J. (2004). Expression and function of RANK in human monocyte chemotaxis. *Arthritis Rheum.*, **50**, 2309-2316.

[22] Takayanagi, H.; Iizuka, H.; Juji, T.; Nakagawa, T.; Yamamoto, A.; Miyazaki, T. Et al. (2000). Involvement of receptor activator of nuclear factor κB ligand/osteoclast differentiation factor in osteoclastogenesis from synoviocytes in rheumatoid arthritis. *Arthritis Rheum.*, **43**, 259-269.

[23] Xing, L.; Schwarz, E. M. & Boyce, B. F. (2005). Osteoclast precursors, RANKL/RANK, and immunology. *Immunol. Rev.*, **208**, 19-29.

[24] McClung, M. R. (2006). Inhibition of RANKL as a treatment for osteoporosis: preclinical and early clinical studies. *Curr. Osteoporos. Rep.*, **4**, 28-33.

[25] Arefieva, T. I.; Kukhtina, N. B.; Antonova, O. A. & Krasnikova, T. L. (2005). MCP-1-stimulated chemotaxis of monocytic and endothelial cells is dependent on activation of different signaling cascades. *Cytokine*, **31**, 439-446.

[26] Jones, G.E.; Prigmore, E.; Calvez, R.; Hogan, C.; Dunn, G.A.; Hirsch, E.; Wymann, M.P. & Ridley, A.J. (2003). Requirement for PI 3-kinase gamma in macrophage migration to MCP-1 and CSF-1. *Exp. Cell. Res.* **290**, 120-131.

[27] Fine, J. S.; Byrnes, H. D.; Zavodny, P. J. & Hipkin, W. (2001). Evaluation of signal transduction pathways in chemoattractant-induced human monocyte chemotaxis. *Inflammation*, **25**, 61-67.

[28] Gerszten, R. E.; Friedrich, E. B.; Matsui, T.; Hung, R. R.; Li, L.; Force, T. & Rosenzweig, A. (2001). Role of phosphoinositide 3-kinase in monocyte recruitment under flow conditions. *J. Biol. Chem.*, **276**, 26846-26851.

[29] Riboldi, E.; Frascaroli, G.; Transidico, P.; Luini, W.; Bernasconi, S.; Mancini, F.; Guglielmotti, A.; Milanese, C.; Pinza, M.; Sozzani, S. & Mantovani, A. (2003). Benzydamine inhibits monocyte migration and MAPK activation induced by chemotactic agonists. *Br. J. Pharmacol.*, **140**, 377-383.

[30] Yen, H.; Zhang, Y.; Penfold, S. & Rollins, B. J. (1997). MCP-1-mediated chemotaxis requires activation of non-overlapping signal transduction pathways. *J. Leukoc. Biol.*, **61**, 529-532.

[31] Cambien, B.; Pomeranz, M.; Millet, M. A.; Rossi, B. & Schmid-Alliana, A. (2001). Signal transduction involved in MCP-1-mediated monocytic transendothelial migration. *Blood*, **97**, 359-366.

[32] Dubois, P. M.; Palmer, D.; Webb, M. L.; Ledbetter, J. A. & Shapiro, L. A. (1996). Early signal transduction by the receptor to the chemokine monocyte chemotactic protein-1 in a murine T cell hybrid. *J. Immunol.*, **156**, 1356-1361.

[33] Laudanna, C.; Campbell, J. J. & Butcher E. C. (1996). Role of rho in chemoattractant-activated leukocyte activation through integrins. *Science*, **271**, 981-983.

[34] Nobes, C. D. & Hall, A. (1995). Rho, Rac and Cdc42 GTPases regulate the assembly of multimolecular focal complexes associated with actin stress fibers, lamellipodia and filopodia. *Cell,* **81**, 53-62.

[35] Weber, K. S. C.; Klickstein, L. B.; Weber P. C.; Weber, C. (1998). Chemokine-induced monocyte transmigration requires cdc42-mediated cytoskeletal changes. *Eur. J. Immunol.,* **28**, 2245-2251.

[36] Nobes, C. D. & Hall, A. (1999). Rho GTPases control polarity, protrusion, and cell adhesion during cell movement. *J. Cell. Biol.,* **144**, 1235-1244.

In: Chemokine Research Trends
Editor: Leah R. Grinwald, pp. 133-145
ISBN: 978-60021-588-9

Chapter 9

THE IMPORTANCE OF THE CCL27-CCR10 CHEMOKINE LIGAND-RECEPTOR INTERACTION IN MELANOMA: AN IMMUNOHISTOCHEMICAL STUDY

Oriana Simonetti**, *Gaia Goteri, Guendalina Lucarini, Graziella Biagini and Annamaria Offidani
Clinica Dermatologica, Anatomia Patologica Dipartimento di Neuroscienze, Istologia Dipartimento di Patologia Molecolare e Terapie Innovative, Università Politecnica delle Marche, Ancona, Italy

ABSTRACT

Malignant melanoma may arise de novo or from potential precursor lesions, such as atypical dysplastic nevi or congenital nevi. The ability of tumour cells to avoid immune surveillance represents a critical step in tumour progression. It has been suggested that the expression of chemokines and chemokine receptors by melanoma cells may be involved in tumour immune escape. Recently ligation of tumor-associated CCR10 with CCL27 has been proposed as a mechanism that promote neoplastic cell survival and reduce immune response in some animal models. Aim of this study was to investigate, by immunohistochemistry, the relation between T immunoreactive cells and the expression of CCR10 and its ligand CCL27 in human melanocytic lesions.

CCL27 and CCR10 expression in melanocytes and the density of CD4+, CD8+ and CD3+ lymphocytes have been assessed in 59 cutaneous melanocytic lesions: 40 malignant melanomas, 10 Spitz nevi, and 9 benign melanocytic nevi.

In malignant melanomas T lymphocyte density was significantly different in the three groups of cases selected by different thickness, showing tendency to decrease from thin melanomas to intermediate and thick ones (Kruskal-Wallis test, $P<0.0005$). CCR10 expression was found both in benign and malignant lesions, being moderate and focal in 15 cases and strong and diffuse in 37. The percentage of expression did not differ between nevi, Spitz nevi and melanomas (Kruskal-Wallis test, $P=0.17$). However, in melanomas with metastatic sentinel lymph node CCR10 mean expression was significantly higher than in cases with negative lymph node (84.1% vs 58.1% in metastatic cases; unpaired t test, $P=0.0281$).

* E-mail address: o.simonetti@univpm.it Tel: +39 71 5963494. Clinica Dermatologica,Ospedali Riuniti, Via Conca, 60020 Torrette (Ancona), Italy

Moreover in melanomas CCR10 shows a tendency to be directly correlated with the Breslow depth (r=0.343; P=0.0298) and inversely with T lymphocyte density (r=-0.363; P=0.0231).

In all benign lesions, CCL27 was negative or faintly expressed by single melanocytes, while in melanomas it was negative in 22 cases, weakly expressed in 16 and strongly and diffusely expressed in 2. When CCR10 and CCL27 expression were evaluated together, CCR10-/CCL27- melanomas tended to have a higher mean density of CD3+ and CD8+ lymphocytes.

Our results confirm that also in human melanomas CCR10 and CCL27 may act to increase the ability of neoplastic cells to grow, invade tissue, disseminate to lymph nodes and to escape the host immune response.

Key words: Melanoma, immunohistochemistry, CCL27, CCR10, host immune response

INTRODUCTION

Cutaneous malignant melanoma, the most aggressive common skin tumor, is characterized by a multi-factorial etiology. Sun exposure and genetic susceptibility have been proposed as major etiological and predisposing factors that somehow explain the reported increased incidence [1-2]. Malignant melanoma may arise de novo or, less frequently, from potential precursor lesions, such as atypical dysplastic nevi or congenital nevi [3-4]. The ability of tumour cells to avoid immune surveillance represents, among other factors, a critical point in tumour progression [5]. The induction of tumor immunity could be viewed as a multistep process that includes: the capture and processing of tumor-associated antigens (TAAs) by dendritic cells via the exogenous endosomal/ lysosomal pathway ; selection and differentiation of antigen-specific T cells as well as non antigen-specific effectors; and finally homing of effector T lymphocytes to the tumor site and elimination of cancer cells [6-8]. It has been suggested that the expression of chemokines and chemokines receptors by melanoma cells may be involved in tumour immune escape [9]. Chemokines are a large family of structurally related, small polypeptide signaling molecules that regulate leukocyte-migration and invasion into tissues by interacting with their specific receptors, which belong to the superfamily of seven–transmenbrane domain G-protein-coupled receptors [10]. The chemokines and their respective receptors are divided into the CXC, CC, C, and CX3C families, based upon the positions of their conserved two N-terminal cys residues. Their genes are clustered on genomic loci, including chromosome 4q12–q13 (CXC acting mainly on neutrophils), 4q21, and 17q11.2 [10].

Although chemokines are classically regarded as proinflammatory mediators, they also exhibit a number of other well characterized functions, having a role in normal tissue development [11], angiogenesis [12] and control of hemopoietic stem and progenitor cell proliferation [13]. They are also involved in cancers in the cellular neoplastic transformation, growth, invasion and homing of metastasis to distant sites [14], and in the host anti-tumor response [5]. Each chemokine affects malignancy by its own specific manners, and the outcome of its activities depends on the combined effects of factor such as: the tumour type and its genetic composition, the factors at the tumor's intimate surroundings, the milieu at remote metastatic sites in which each tumor preferentially metastasizes, and the diversity of other chemokine and chemokine receptors expressed by tumor cells, as well as by host cells [15]. Chemokine can affect directly tumor cells, mainly promoting but also potentially suppressing cell characteristics and functions involved in tumor development and progression

[15]. It has been hypothesized that migration of tumor cells in specific organs may be determined by the chemokine receptors the cells express and by the chemokines expressed in the target organs [9]. Specifically, melanoma cells were found to express high levels of CXCR4, CCR7, and CCR10 mRNA as compared with normal melanocytes [5]. Recent investigations demonstrated in murine melanoma that B16 cells overexpressing CCR10 resisted host immune responses and readily formed tumors [16]. CCR10 is a specific receptor of the skin-related chemokine CCL27 found to be a selective chemo-attractant for cutaneous-associated Ag-positive T and Langerhans cells [17] and could play a role in T cell homing in inflammed skin [18,19].

The role of the CCR10/CCL27 interaction in melanomas is unclear at present. It is possible to suppose that CCL27 in melanomas can induce anti-tumoural immunity, as described by Guo et al. in a murine lung carcinoma model, where CCL27-mediated tumor regression was obtained by an efficient induction of antitumor immunity [20]; CCL27 could suppress tumor growth probably due to the local recruitment of T cells and natural killer cells [21]. However, recent publications on murine melanoma have suggested that ligation of tumor-associated CCR10 with CCL27 may be a mechanism to promote cancer cell survival and reduce immune response in animal models [16].

Aim of this study was to investigate the relation between T immunoreactive cells and the expression of CCR10 and its ligand CCL27 in human melanoma tissues, with respect to tumour thickness, invasion levels and lymph node metastasis.

Material and Methods

Fifty-nine cases of cutaneous melanocytic lesions consecutively diagnosed at the Institute of Anatomic Pathology of Polytechnic University of Marche between 2002-2004 were included in this study. The original slides were retrieved from the Archive and the histological diagnosis was first reavaluated and confirmed in all cases according to the classical criteria [22]. Cases included in the study were 40 primary cutaneous human melanomas and 19 benign lesions (10 spindle and epithelioid cell Spitz nevi, 9 ordinary benign melanocytic nevi, either compound or predominantly intradermal).. Melanomas were invasive in all case: Clark's levels were II-III in 16 cases, and IV-V in 24 cases. Fifteen had a Breslow thickness< 1.0 mm; twelve of 1.1-4.0 mm; thirteen > 4.1 mm. None of them presented regression areas. Clinical history was obtained by original medical charts at the Clinic of Dermatology. Patients with melanomas were 17 males and 23 females. Twenty-four patients underwent examination of sentinel lymph node: except for one patient who had a thin melanoma, all the others had a thick (11 cases) or an intermediate melanoma (12 cases). Ten out of 24 patients (42%) were positive for metastasis (6 females, 4 males; 6 intermediate and 4 thick melanomas).

Immunohistochemistry

Immunohistochemistry was performed at the Institute of Histology for evaluation of CCR10 and CCL27 expression, and at the Institute of Anatomic Pathology for evaluation of lymphocytic population density. For each case the most representative paraffin inclusion was selected, from which conventional 6 μm thick histological sections were obtained with a microtome and mounted on slides pre-treated with poly-l-lysine (Sigma Chemicals, St Louis, MO, USA). The slides were then deparaffined and rehydrated in a gradient of ethanol and xylene. Antigenic retrieval was performed with two different methods: sections were incubated with a solution of TUF (Target Unmasking Fluid; Alexis, Biochemical) at 90° C for 10 min before staining with anti-CCR10 and CCL27 and with a solution of EDTA 0.01M at pH 8.0 at 95-98° C for 30 minutes before staining with anti-CD3, -CD4 and -CD8. After washing in H2O and then in Tris buffer solution the sections were incubated with the following monoclonal antibodies: mouse anti-CCL27 (dil. 1:200, R&D Systems, Minneapolis, MN), policlonal antibody anti-CCR10 (Alexis-Italia, Vinci, Firenze, Italy), mouse anti-CD4 (1F6, dil. 1:40, Novocastra Laboratories Ltd, Newcastle, UK), mouse anti-CD8 (4B11, dil. 1:40, Novocastra Laboratories Ltd, Newcastle, UK) and rabbit anti-CD3 (dil. 1:100, Dakocytomation, Milano, Italy). Incubations with the antibodies were carried out in a humidified atmosphere at room temperature overnight at 4° C for anti-CCL27 and anti-CCR10, and for 1 hour for anti-CD3, -CD4, -CD8. The antibodies were linked to an alkaline phosphatase anti-alkaline phosphatase (APAAP) catalyzed reaction (Kit Dakocytomation, Milano, Italy) being washed roughly with PBS between all stages in the procedure. For visualization of the antibody sections were incubated with the substrate Fas red (Dakocytomation, Milano, Italy), yielding a red end product at the site of the target antigen. The sections were finally counterstained with Mayer's haematoxylin, dehydrated and mounted in glycerol-gelatin (Dakocytomation, Milano, Italy). Each series included appropriate negative control sections omitting the primary antibody reaction.

Immunostainings for CCR10 and CCL27 were also evaluated with a light microscope by scanning the entire melanocytic lesions: epidermal and dermal positive melanocytes were counted together and their amount was expressed as percentage of positive cells.

T lymphocyte subsets were evaluated with a light microscope by scanning the entire melanocytic lesions: intra-lesional and peri-lesional positive lymphocytes were counted together and their amount was expressed as number of positive cells per high power field (HPF).

Statistical Analysis

Statistical analysis was performed in order to compare the CCR10 and CCL27 expression in melanocytes in different histologic lesions (benign nevi, Spitz nevi, malignant melanomas); in particular, in melanomas these variables were correlated with CD3+, CD4+ and CD8+ T lymphocyte density and analysed with respect to biologic characteristics of the tumors, like the depth of infiltration, measured on histologic slides, Clark's level and metastasis in sentinel lymph node. For statistical analysis, two categories of Clark's level were considered by grouping together cases with Clark's level II and III and cases with level IV and V; cases with Breslow depth <1.0 mm were grouped in the category of thin melanomas, cases with Breslow thickness >1.1 and <4.0mm were grouped in the category of intermediate melanomas, whereas cases with Breslow thickness >4.0 were considered thick melanomas.

Data were summarised by mean and range for continuous variables. Differences between mean values in different categories were analysed by unpaired Student t test and non parametric Kruskal-Wallis test. Correlations between continuous variables were analysed by means of the Spearman correlation test. All statistical analysis was performed by SPSS statistical package (SPPS Inc. Chicago, IL).

RESULTS

Data regarding melanocytic expression of CCR10 and CCL27, and the density of T lymphocytes and helper or suppressor subsets in the different diagnostic categories, are reported in table 1and 2.

Table 1. CCR10+ and CCL27+ melanocytic expression in benign nevi and malignant melanomas.

Histology	CCR10+ % positive cells (mean±SD)	CCL27+ % positive cells (mean±SD)
Benign nevi	42.8±30.1	0.31±0.6
Spitz' nevi	56.9±28.9	4.9±10.7
Melanomas	62.0±33.5	12.6±20.6
II-III Clark's level	51.5±37.4	7.6±10.4
IV-V Clark's level	69.0±29.3	16.0±24.9
Breslow <1.0mm	46.7±36.8	7.6±10.7
Breslow >1.0<4.0mm	61.9±34.8	17.2±31.6
Breslow > 4.0mm	79.7±17.8	14.1±16.6
Negative sentinel lymph node	58.1±33.7	14.4±23.4
Positive sentinel lymph node	84.1±10.1	18.8±27.5

SD: standard deviation.

CCR10 and CCL27 Immunostaining

CCR10 immunostaining was not found in 7 cases (1 benign nevus, 1 Spitz nevus and 5 melanomas). The other cases showed a variable cytoplasmic staining, ranging from moderate and focal in 15 cases (4 benign nevi, 4 Spitz nevi and 7 melanomas), to strong and diffuse in 37 (4 benign nevi, 5 Spitz and 28 melanomas; fig. 1a, 1b). The percentage of expression did not differ between nevi, Spitz nevi and melanomas (Kruskal-Wallis, P=0.17). However, in melanomas there was a correlation between high CCR10 expression and metastasis in sentinel lymph node (58.1% in negative cases vs 84.1% in metastatic cases; unpaired t test, P=0.0281). Moreover, CCR10 expression in malignant melanocytes was correlated directly with the Breslow depth (r=0.343; P=0.0298).

CCL27 was expressed in the cytoplasms of epidermal keratinocytes in all lesions. In all benign lesions, CCL27 was negative or faintly expressed by single melanocytes, except in one case of Spitz nevus, which showed a moderate and focal expression. In melanomas CCL27 was negative in 22 cases, weakly to moderately expressed in 16 (fig. 1c), and strongly and diffusely expressed in 2 (fig. 1d).

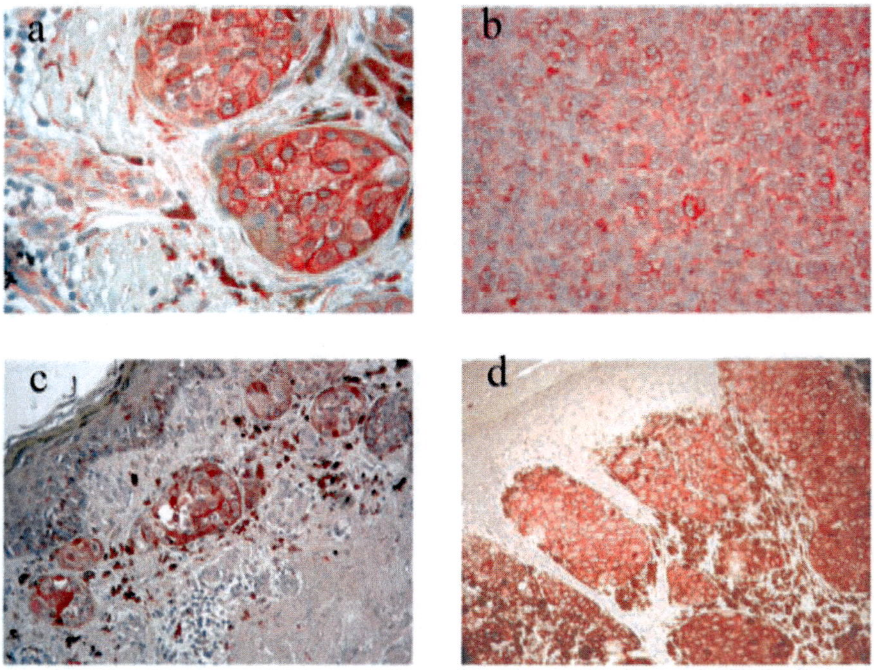

Figure 1. Strong and diffuse immunohistochemical expression of CCR10 in thin (a) (APAAP, x 400) and in thick (b) melanomas (APAAP, x 200). Immunohistochemical CCL27 staining appears moderate in thin melanoma (c) (APAAP, x 100) and/or strong in intermediate melanoma (d) (APAAP, x100).

Table 2. CD3+, CD4+, CD8+ lymphocytes density in benign nevi and malignant melanomas

Histology	Cases (n. cases)	Lymphocyte density (number of cells/HPF) mean ± SD					
		CD4+	p	CD8+	p	CD3+	p
Melanocytic lesion	BENIGN NEVI(9)	1.9±2.9		6.1±4.2		8.1±5.9	
	Spitz'nevi (10)	7.9±5.9		9.1±7.9		17.0±12.3	
	[c]Melanomas (40)	13.6±11.8	*(vs a and b)	17.2±12.3	*(vs a and b)	31.0±20.6	*(vs a and b)
Clark's level	II-III(16)	21.2±12.9	**(a vs b)	24.4±14.2	**(a vs b)	45.6±20.1	**(a vs b)
	IV-V (24)	7.9±6.8		12.3±7.6		20.2±13.3	
Breslow	[a]<1.0mm(1)	20.9±13.3		23.1±13.9		44.0±19.9	
	>1.0<4mm)	12.9±8.0		18.2±10.6		30.0±17.1	
	>4.0mm(13)	4.4±4.0	*(vs a and b)	9.5±6.9	*(vs a and b)	13.6±10.2	*(vs a and b)
Sentinel lymph node	Negative(4)	8.2±7.3		13.7±8.1		21.9±14.1	
	Positive(10)	10.1±8.3	NS	15.1±11.8	NS	25.5±20.0	NS

SD: standard deviation; HPF: High power field; *: P< 0.0005, Kruskal Wallis test; **: P<0.0001, Unpaired T, NS: not significant

T Lymphocyte Density

T lymphocyte density was correlated with the histologic type of lesions. It was indeed found higher in malignant melanomas compared to benign nevi and Spitz nevi (Kruskal-Wallis test: P<0.0005, Fig.2).

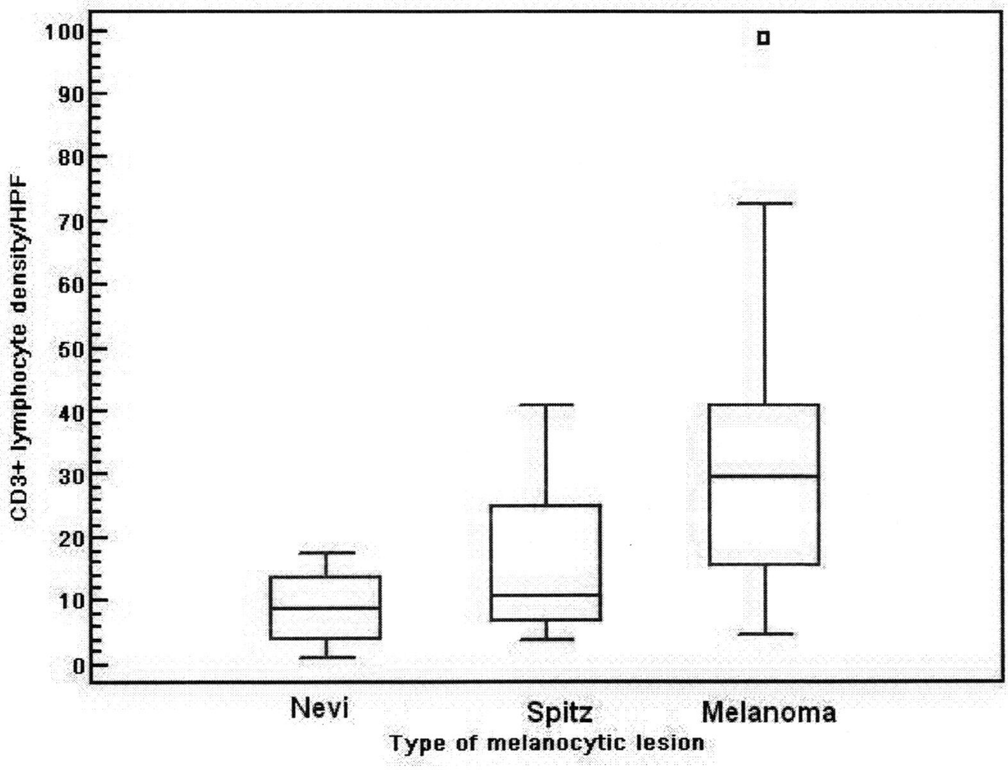

Figure 2. The graph shows a correlation between T lymphocyte density and the histologic type of lesions. CD3+ cell density was significantly higher in malignant melanomas compared to benign nevi and Spitz nevi (Kruskal-Wallis test: P<0.0005).

The increase of T lymphocytes was not limited to a particular subset of T lymphocytes, because both CD4+ and CD8+ cells showed the same trend (Kruskal-Wallis test: P<0.0005 for both). In malignant melanomas, T lymphocyte density was found inversely correlated to the depth of infiltration (r=-0.596, P<0.0001): melanomas cases with II-III Clark level showed a higher T lymphocyte mean density compared to cases with IV-V Clark's level (unpaired t test: P<0.0001, Fig 3).T lymphocyte density was significantly higher in thin melanomas compared to intermediate and thick melanomas (Kruskal-Wallis test, P<0.0005, fig. 4). T lymphocyte density did not correlate with metastasis in sentinel lymph node (unpaired t test, P=0.96). T lymphocyte density was inversely correlated with CCR10 expression (r=-0.363; P=0.0231) and CCL27 expression in malignant melanocytes (r=-0.333; P=0.0381).

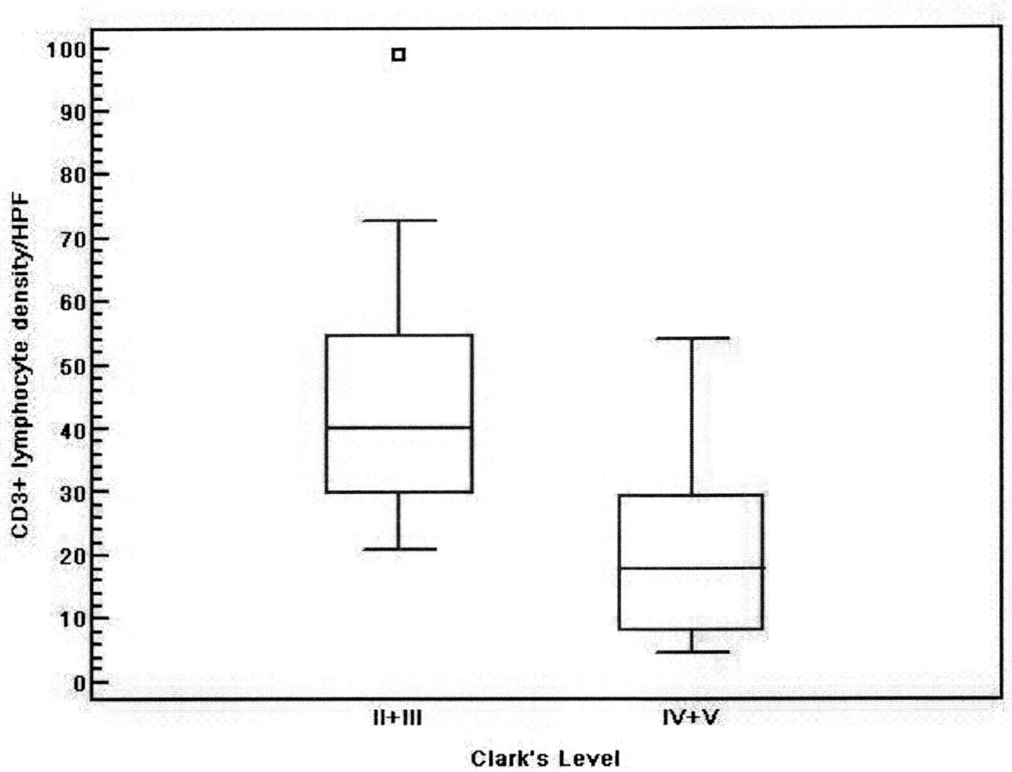

Figure 3. T lymphocyte density is related to Clark level in malignant melanocytes. Melanomas cases with II-III Clark level showed a higher T lymphocyte mean density compared to cases with IV-V Clark's level (unpaired t test: $P<0.0001$).

Ten cases of melanomas expressed both CCR10 and CCL27 in more than 20% of the cells (+/+ phenotype), 23 cases expressed only CCR10 in more than 20% of the cells (+/- phenotype), whereas 7 cases were negative or expressed both antigens in less than 20% of the cells (-/- phenotype). The T lymphocyte density was significantly different according to the phenotype: cases with a double negative phenotype tended to have a higher mean density of CD3+ (47.16/HPF) and CD8+ lymphocytes (24.89/HPF) compared to cases showing expression of CCR10 alone (CD3+: 31.51/HPF; fig. 4; CD8+: 18.72/HPF; fig. 5) or CCR10 and CCL27 (CD3+: 18.01/HPF; CD8+: 8.87/HPF; Kruskal-Wallis for CD3+ density: $p=0.0154$; for CD8+ density: $p=0.009$). The tumor thickness tended to be different according to the phenotype, but without reaching a statistical significance: cases with a double negative phenotype tended to have a lower tumor thickness (0.90 mm) compared to cases showing expression of CCR10 alone (3.17 mm) or both CCR10 and CCL27 (3.95 mm Kruskal-Wallis: $p=0.13$).

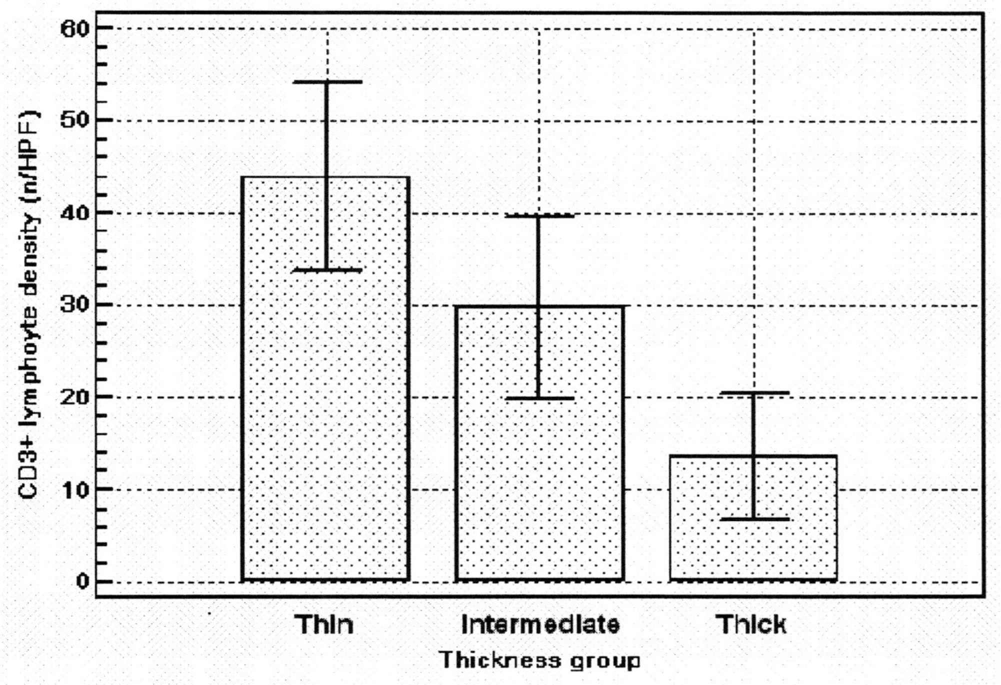

Figure 4. The graph shows that the CD3+ lymphocyte density tends to reduce inversely to the tumour thickness in malignant melanomas: mean values and standard deviations were, respectively, 44.0±19.9 in thin melanomas, 31.1±17.1 in intermediate, 13.6±10.2 in thick ones. Differences were statistically significant (Kruskal-Wallis, $P<0.0005$) .

DISCUSSION

Malignant melanoma has always been regarded as an immunogenic tumour, as regression zones within tumoral lesions can be observed frequently together with a dense infiltration of T helper cells [23], that results from recognition of tumour-associated antigens either on antigen-presenting cells or on the surface of tumour cells by T lymphocytes [24-25].

Seminal observations has shown that tumour infiltrating lymphocytes represent a significant prognostic indicator of survival especially in cases with a vertical growth [26-27]. It is still an unsolved question why cancer patients can naturally develop antitumor T cell immune responses that may fail to effectively prevent tumor progression and recent data indicate that the presence of clonally expanded T cells in melanomas may represent subsets with different functions. Tumoral lymphoid population is in fact composed not only of CD8+ cytotoxic lymphocytes or CD4+/CD25^{-} cells, specific for tumoral antigens, but also of regulatory T cells with a CD4+/CD25^{high} phenotype and inhibiting local antitumor response [28-29]. In our study on melanomas with a wide range of level of infiltration and with both horizontal and vertical growth, we found an inverse correlation between T lymphocyte density and tumour thickness, according to other investigators [30], suggesting that melanomas with a greater thickness show a reduction of immune competent cells regardless of lymphocyte specific function and phenotype. We studied the interaction between

CCL27/CCR10 and T cell populations, considered a recent mechanism that explains T cell trafficking in the skin. Notably, CCR10 is expressed in normal cellular components of the skin including melanocytes, dermal fibroblasts and microvascular endothelial cells, T lymphocytes and Langerhans cells [17]. CCR10 expression has been already observed in melanomas [16], where its role is still unclear. It could mediate tumour resistance to Fas-mediated apoptosis, which constitutes the major mechanism by which immune cells, natural killer and cutaneous T lymphocytes, expressing the Fas-ligand, kill Fas-positive tumour cells [31]. Murakami et al. showed that B16 murine melanoma cells CCR10+ also expressed, *in vitro* , low levels of surface Fas protein and were more resistant to Fas-dependent cell death [16]. CCL27 protein, a skin specific homeostatic chemokine, is highly expressed in epidermal basal keratinocytes, immobilized on ECM or displayed on the surface of dermal endothelial, where it regulates the homing of memory T cells into the skin [17-19]. CCL27 expression has never been described in patients affected by malignant melanoma. Moreover studies *in vitro* have shown conflicting results on the possible role of CCL27 in tumours, showing that either CCL27 could suppress tumor growth probably by local recruitment of T cells and natural killer cells [21] or, conversely, increase the resistance of CCR10+ murine melanoma cells to Fas-mediated apoptosis, resulting in markedly decreased killing of target cells and in a reduction of T cells [16].

Our results suggest that the aggressive behaviour and the immune escape in melanomas could involve CCR10 together with its ligand CCL27. CCR10 immunopositivity of melanocytes was present both in benign (89%) and malignant (84%) melanocytic lesions; moreover we observed a strong CCR10+ immunostaining in 70% of malignant melanomas and only in 47% of benign nevi. From our results it can be inferred that malignant neoplastic cells are characterised by a CCR10 overexpression. There are evidences that CCR10 expression is induced in melanoma cells by TNF-α and IL-1β [17] and a number of studies have shown that melanoma can produce several growth factors and cytokines, including IL-1β and TNF-α [32-33]; therefore, these molecules may lead to CCR10 overexpression in melanoma cells. In melanomas we observed a direct correlation of CCR10 with tumour thickness and metastasis in sentinel lymph nodes, and an inverse correlation with T lymphocytic density. It follows that CCR10 expression might be related to an aggressive behaviour of melanoma and to its ability to escape immune control. Experiments on animals have already shown that CCR10 expression in B16 melanoma cells is associated with a striking increase in metastasis to the draining lymph node after injection into the footpads of mice [16]. In human malignant melanomas, about 60% of metastases occur in regional lymph nodes [34], thus lymph node metastasis status represent an important predictor of survival for this disease [27]. An important role in determining the aggressive behaviour and the immune escape of melanomas can also be played by the CCR10-ligand CCL27. We found CCL27 expression in the cytoplasm not only of basal keratinocytes, but also of malignant melanocytes, in 45% of melanoma cases. Benign melanocytes from nevi were all negative or faintly CCL27 positive, except one case of Spitz. CCL27, produced in the epidermis and then secreted into the papillary dermis, could be immobilized on the extracellular matrix and subsequently displayed in the cytoplasm of melanocytic cells following transcytosis. This mechanism has been observed to occur normally in endothelial cells of superficial dermal plexus by Homey et al. [18] and probably requires CCR10 expression. In fact Gortz et al. [35] showed that exogenous mature CCL27 may be able to enter target cells by interacting with its receptor, CCR10, in a paracrine manner. Indeed, we observed that melanoma samples negative for CCR10 receptor, did not even show CCL27 chemokine immunostaining. The capability to internalize the CCL27 chemokine might be a feature of malignant melanocyte, because it was observed more frequently in melanomas than in benign lesions: only one Spitz nevus was found to be immunostained for CCL27. Interestingly, a difference was observed between this benign lesion and the melanoma cases in relation to immune cell density. Infact, while the Spitz nevus showed an abundance of T cells, CCL27+ melanomas were associated with a significant lower CD3+ and CD8+ mean density compared to CCL27- melanomas, suggesting that CCR10/CCL27 ligation could make melanoma cells less susceptible to the host antitumor response, accordingly to other authors in animal models [16]. Moreover, the coexpression of CCR10 and CCL27 was correlated with a higher tumour thickness, suggesting that ligation of CCR10 by CCL27 in melanoma cells could be associated with an advantage in tumour growth [16]. Since CCL27 can be regarded as a cellular response that recruit lymphocytes to the tumor site, the CCL27 internalization in neoplastic melanocytes expressing CCR10, by the CCL27/CCR10 interaction, could lead to a sequestering of the chemokine as an attempt to withdraw lymphocyte recruitment. Different sequelae can occur in benign and malignant cells when CCL27 is internalized and then translocated to the nucleus where it can mediate a number of biological effects. One consequence might be an increase of migratory competence by inducing actin cytoskeletan relaxation, as recently reported [35].

In conclusion, our preliminary results confirm that also in human melanomas CCR10 may act in increasing the ability of neoplastic cells to grow, invade tissue, disseminate to lymph nodes and to escape the host immune response. We suggest that co-expression of CCL27 and CCR10 in neoplastic cells could be linked with poor prognosis in melanoma patients. A better understanding of the CCR10/CCL27 interaction in melanomas could have future potential therapeutic applications by inhibiting the function of the chemokine and its receptor and subsequently decreasing the ability of neoplastic cells to grow, disseminate to other organs and modulate the host immune response mediated by T lymphocytes against the tumour cells.

REFERENCES

[1] Nordlund, JJ; Boissy, RE; Hearing, VJ et al., eds. The pigmentary system: Physiology and Pathophysiology. New York, NY. *Oxford University Press*, 1988.

[2] Lejeune, FJ; Chaudhuri, PK; Das Gupta, PK; eds. Malignant Melanoma and surgical management. New York, *NY: McGrow-Hill*, 1994.

[3] Clark, W ;Elder, DE ; Guerry, D ; et al.. A study of tumor progression: the precursor lesions of superficial spreading and nodular melanoma. *Hum Pathol*, 1984 15: 1147-65.

[4] Herlin, M; Clark, WH; Rodeck, U; et al. Biology of disease: Biology of tumor progression in human melanocytes. *Lab Invest*, 1987 56: 461-74.

[5] Payne, AS; Cornelius, LA. The role of chemokines in melanoma tumor growth and metastasis. *J Invest Dermatol*, 2002 118: 915-22.

[6] Pawelec, G; Zeuthen, J; Kiessling, R. Escape from host-antitumor immunity. *Crit Rev Oncog* 1997 8: 111-141.

[7] Pardoll, DM. Cancer vaccines. *Nat Med* 1998 4: 525-531.

[8] Sogn, JA. Tumor immunology: the glass is half full. *Immunity* 1998 9: 757-763.

[9] Muller, A; Homey, B; Soto, H; et al. Involvement of chemokine receptors in breast cancer metastasis. *Nature* , 2001 410: 50-56.

[10] Zlotnik, A; Yoshie, O. Chemokines: a new classification system and their role in immunity. *Immunity*, 2000 12: 121-127.

[11] Nagasawa, T; Hirota, S; Tachibana, K; et al. Defects of B-cell lymphopoiesis and bone-marrow myelopoiesis in mice lacking the CXC chemokine PBSF/SDF-1. *Nature*, 1996 382: 635-81.

[12] Strieter, RM; Di Giovine, B; Polverini, S; et al. C-X-C chemokines and lung cancer angiogenesis. In Rollins, BJ. ed. Chemokines and cancer, *Humana Clifton NJ*, 1999; 143.

[13] Graham, GJ. Growth inhibitors in haemopoiesis and leukemogenesis. *Ballieres Clin Haematol*, 1997 10: 539-559.

[14] Strieter, RM; Polverini, S; Kunkel, SL; et al. The functional role of ELR motif in C-X-C chemokine mediated angiogenesis. *J Biol Chem*, 1995 270: 1083-86.

[15] Ben-Baruch A. The multifactorial roles of chemokines in malignancy. *Cancer Metastasis Rev*, 2006, 3.

[16] Murakami, T; Cardones, AR; Finkelstein, SE; et al. Immune evasion by murine melanoma mediated through CC chemokine receptor-10. *J Exp Med*, 2003 198: 1337-47.

[17] Homey, B; Wang, W; Soto, H; et al. Cutting Edge: the orphan chemokine receptor G protein-coupled receptor-2 (GPR-2, CCR10) binds the skin associated chemokine CCL27 (CTAK/ALP/ILC). *J Immunol*, 2000b 164: 3465-70.

[18] Homey, B ; Alenius, H ; Muller, A ; et al. CCL27-CCR10 interactions regulate T cell mediated skin inflammation. *Nat Med*, 2002 8: 157-165.

[19] Morales, J; Homey, B; Vicari, AP; et al. CTHCK, a skin associated chemokine that preferentially attracts skin homing-memory T cells. *Proc Natl Acad Sci USA* , 1999 96: 14470-5.

[20] Guo, J; Wang, B; Zhang, M; et al. Macrophage-derived chemokine gene transfer results in tumor regression in murine lung carcinoma model through efficient induction of antitumor immunity. *Gene therapy*, 2002 9: 793-803.

[21] Okada, N; Gao, J-Q; Sasaki, A; et al. Anti-tumor activity of chemokine is affected by both kinds of tumors and the activation state of the host's immune system: implications for chemokine-based cancer immunotherapy. *Biochem Biophys Res Com*, 2004 317: 68-76.

[22] Lever, WF; Schaumburg-Lever, G . In: Histopathology of the Skin. 7th edn. Philadelphia: *J.B.Lippincott*,; 1990, 756-805.

[23] Nathanson L. Spontaneous regression of malignant melanoma: a review of the literature on incidence, clinical features, and possible mechanisms. *Natl Cancer Inst Monogr*, Nov; 1976 44:67-76.

[24] Boon, T; Coulie, PG; Van den Eynde, B. Tumor antigens recognized by T cells. *Immunol* Today, 1997 18: 267-8.

[25] Romero, P; Dunbar, PR; Valori, D; et al. Ex vivo staining of metastatic lymph nodes by class I major histocompatibility complex tetramersmreveals high number of antigen experienced tumor-specific cytolytic T lymphocytes. *J Exp Med*, 1998 188:1641-50.

[26] Clemente, CG ; Mihm, MC ; Bufalino, R ; et al. Prognostic value of tumor infiltrating lymphocytes in the vertical growth phase of primary cutaneous melanoma. *Cancer*, 1996 77:1303-10.

[27] Clark, WH Jr ; Elder, DE ; Guerry, DP IV ; et al. Model predicting survival in stage I melanoma, based on tumor progression. *J Natl Cancer* Inst, 1989 81:1893-904.

[28] Viguier, M; Lemaitre, F; Verola, O; et al. Foxp3 expressing $CD4^+/CD25^{high}$ regulatory T cells are overpresented in human metastatic melanoma lymph nodes and inhibit the function of infiltrating T cells. *J Immunol*, 2004 173:1444-53.

[29] Bernsen, MR; Dipestra, JHS; Van Mil, P; et al. Presence and localization of T-cell subsets in relation to melanocyte differentiation antigen expression and tumour regression as assessed by immunohistochemistry and molecular analysis of microdissected T cells. *J Pathol*, 2004 202: 70-79.

[30] Lyle, S; Salhany, KE; Elder, DE. TIA-1 positive tumor-infiltrating lymphocytes in nevi and melanomas. *Mod Pathol*, 2000 13:52-5.

[31] Strand, S; Hofmann, WJ; Hug, H; et al. Lymphocyte apoptosis induced by CD95 (APO-1/Fas) ligand-expressing tumor cells: a mechanism of immune evasion ? *Nat Med*, 1996 2: 1361-66.

[32] Lazár-Molnár, E; Hegyesi, H; Tóth, S; et al. Autocrine and paracrine regulation by cytokines and growth factors in melanoma. *Cytokine*, 2000 12: 547-54.

[33] Slominski, A; Wortsman, J; Carlson, AJ; et al. Malignant melanoma. Arch Pathol *Lab Med*, 2001 125(10):1295-306.

[34] McCarthy, WH; Shaw, HM; Thompson, JF; et al. Time and frequency of recurrence of cutaneous stage I malignant melanoma with guidelines for follow-up study. *Surg Gynecol Obstet*, 1988 166: 497-502.

[35] Gortz, A; Nibbs, RJB; McLean, P; et al. The Chemokie Eskine/CCL27 displays novel modes of intracrine and paracrine function. *J Immunol*, 2002 169: 1387-94.

In: Chemokine Research Trends
Editor: Leah R. Grinwald, pp. 147-158
ISBN: 978-60021-588-9

Chapter 10

CROSS TALK BETWEEN CHEMOKINE RECEPTOR CXCR4 AND ONCOPROTEINS IN TUMOR PROGRESSION AND DISSEMINATION/METASTASIS

Seiji Fukuda* and Louis M. Pelus

Department of Microbiology and Immunology and the Walther Oncology Center, Indiana University School of Medicine, and the Walther Cancer Institute, Indianapolis, IN 46202

ABSTRACT

The chemokine SDF1(CXCL12) and its receptor CXCR4 play pivotal roles in survival and metastasis of cancer cells. Recent studies have identified functional cross talk between the SDF1/CXCR4 pathway and several oncogenes that regulate migration, invasion, survival, trafficking and metastasis of transformed cells.

The receptor tyrosine kinase, HER2, which is frequently over-expressed in breast cancer cells and associated with an adverse prognosis, up-regulates CXCR4 expression and migration and invasiveness of breast cancer cells in response to SDF1. Moreover, CXCR4 is required for HER2-mediated metastasis to lung in vivo. Conversely, SDF1 transactivates HER2 in breast cancer cells. These results implicate the interaction of HER2 and SDF1/CXCR4 signaling pathways in the regulation of breast cancer progression and metastasis.

Similar to metastasis by solid tumor cells, leukemia cells leave the marrow and accumulate in the spleen and blood and often infiltrate extramedullary sites, suggesting that signaling events that regulate migration, homing and mobilization may be deregulated by the genes associated with hematological malignancies. Elevated CXCR4 expression is associated with Internal Tandem Duplication of the tyrosine kinase Flt3 (ITD-Flt3) and is also associated with poor prognosis in patients with AML. SDF1/CXCR4 and ITD-Flt3 functionally interact to enhance hematopoietic cell survival and migration. Furthermore, ITD-Flt3 decreases adhesion of hematopoietic cells to VCAM-1 and facilitates their accumulation in the spleen, suggesting that ITD-Flt3 modulates trafficking of transformed hematopoietic cells. In contrast to ITD-Flt3, the oncogene BCR/ABL, which is found in patients with Chronic Myeloid Leukemia (CML), down regulates CXCR4 expression and migration to SDF1, which may contribute to defects in retention and homing of CML cells.

Understanding the mechanisms by which oncoproteins regulate SDF1/CXCR4 signaling and function will facilitate development of novel strategies to block metastasis and eliminate

* E-mail address: sfukuda@iupui.edu or sfukuda@med.shimane-u.ac.jp, Phone: 317-278-9944, Fax: 317-274-7592

development of resistance to apoptosis characteristic of cancer cells and leukemia cells. This chapter will summarize the recent evidence on signaling cross talk between oncoproteins and SDF/CXCR4 pathways that may contribute to the dissemination, metastasis and enhanced survival of cancer cells and leukemia cells.

Physiological Role of SDF1 and CXCR4

The chemokine stromal derived factor-1 (SDF1) (CXCL12) is expressed in spleen, liver, lung, kidney, thymus, brain [1], stromal cells [2] and osteoblasts in bone marrow [3;4] and regulates development of hematopoietic cells, immune cells, blood vessels, heart and brain [5;6]. Interaction between SDF1 and its unique receptor CXCR4 is believed to play an important role in these processes. CXCR4 was also found to serve as a co-receptor for HIV virus entry into $CD4^+$ T-cells [7]. One of the major roles of SDF1/CXCR4 in normal hematopoiesis is to regulate homing, retention, and survival of primitive hematopoietic stem and progenitor cells [8-12]. SDF1 can attract primitive hematopoietic cells that express CXCR4 to the marrow microenvironment [8;10], while disruption of SDF1/CXCR4 interaction within marrow can under appropriate circumstances facilitate their mobilization to the peripheral circulation [9;12]. SDF1 can activate the adhesion receptors, very late antigen-4 (VLA-4) and lymphocyte function associated antigen-1 (LFA-1) on HSPC, which can contribute to the homing process [13]. SDF1 also enhances survival or proliferation of normal hematopoietic progenitor cells [14, 15] and regulates development of B-cells [16;17]. Although redundancy between chemokines and their receptors occurs for the majority of chemokine/chemokine receptor interactions, CXCR4 is the only receptor for SDF1. Consistent with this finding, the phenotype of mice in which SDF1 or CXCR4 is functionally deleted display identical, lethal phenotype [5;18].

Role of SDF1/CXCR4 in Hematological Malignancies and Solid Tumors

In addition to the physiological roles of SDF1 and CXCR4, CXCR4 is expressed in most hematological malignancies including chronic lymphocytic leukemia, chronic myeloid leukemia (CML) [19], acute lymphoblastic leukemia (ALL) [20;21], acute myeloid leukemia (AML) [21-23], B-cell lymphoma [19;24] T-cell non Hodgkin lymphoma (NHL) [25] and multiple myeloma [26]. Normal cells expressing functional CXCR4 migrate in response to SDF1 and home to the hematopoietic microenvironment where SDF1 is expressed and supports their survival [27]. Leukemia cells also migrate in response to SDF1 [21;28]. It has been suggested that targeting CXCR4 signaling pathways may be an important therapeutic strategy for ALL [29], CLL [30] and NHL [31]. High expression of CXCR4 predicts extramedullary organ infiltration in childhood ALL [32] and elevated CXCR4 expression is normally associated with poor prognosis of patients with AML [22;33]. The SDF1/CXCR4 axis may enhance adhesion of AML cells to stromal cells, which can be an additional factor that may increase resistance to treatment, since adhesion of AML cells to stromal fibronectin increases resistance to cytotoxic drugs and is associated with minimal residual disease [34].

Elevated CXCR4 is expressed by many solid tumors and associated with disease progression. CXCR4 signaling is involved in migration and survival of cancers, including breast [35], lung [36;37], prostate [38] and ovarian cancer [39], neuroblastoma [40], melanoma [41] and brain tumors [42]. Recent evidence suggests that SDF1/CXCR4 signaling plays a central role in metastasis of solid tumors to the bone marrow or other tissues where SDF1 is expressed [35;40] and serves as a survival factor for various solid tumor or leukemia cells [42-44]. Furthermore, SDF1 can recruit endothelial progenitor cells to the tumor microenvironment to facilitate neo-vascularization, which is required for continued tumor growth [43]. A growing body of evidence indicates that SDF1/CXCR4 signaling and oncogenic proteins, such as BCR/ABL in chronic myeloid leukemia (CML) [37;45-47], HER2 in breast cancer [48;49] and Internal Tandem Duplication (ITD)-Flt3 in AML [50] functionally interact to modulate the aberrant biological behavior of malignant cells. Their concomitant expression with CXCR4 is often associated with poor patient prognosis [22;49].

Functional Cross Talk Between SDF1/CXCR4 and HER2 that Modulates Migration, Adhesion and Metastasis of Breast Cancer Cells

Metastasis is one of the major causes of mortality in almost all cancers. Recent publications provide evidence that SDF1/CXCR4 plays a critical role in regulating metastasis of tumor cells [35;40]. SDF1 is expressed in multiple organs, such as bone marrow, lung, liver and brain [1], which are the primary target sites for cancer metastasis. HER2 (ErbB2) is overexpressed in many types of cancers, including ~30% of breast cancer patients and is associated with poor prognosis. Li et al. demonstrate a functional link between HER2 and CXCR4 and that CXCR4 overexpression plays a central role in breast cancer metastasis induced by HER2 [49]. HER2 increases expression of CXCR4 in breast cancer cells through activating the PI3 kinase/Akt/mTOR pathway and inhibiting SDF-mediated CXCR4 degradation. Conversely, inhibiting HER2 expression results in a reduction in CXCR4 expression. Increased CXCR4 expression is essential for cell invasion, migration, and adhesion induced by SDF1. The enhanced invasion induced by SDF1 is inhibited in HER2 positive BT474 cells by CXCR4 inhibitory antibodies, CXCR4 RNAi or an inhibitor for HER2, indicating a functional link between HER2 and CXCR4 in breast cancer cell invasion. More importantly, the study shows that CXCR4 expression on breast cancer cells is required for their metastasis to lung in a mouse lung cancer model. Conversely, SDF1/CXCR4 signaling transactivates HER2 in breast cancer cells through Src kinase activation, suggesting that SDF1 can increase HER2 signaling [48]. Consistent with these findings, concomitant expression of HER2 and CXCR4 was observed in 22% of patients with breast tumors and correlated with a poor overall survival rate [49]. These results indicate that SDF1/CXCR4 and HER2 signaling regulate each other and contribute to the invasion and metastasis potential of breast cancer cells.

BCR/ABL Alters Signaling and Chemotactic Response Mediated by SDF1/CXCR4

Chronic myeloid leukemia (CML) is caused by the BCR/ABL oncogene, a result of a reciprocal translocation between chromosmes 9 and 22. It is found in cells of 95% of patients with CML and encodes a cytoplasmic protein with constitutive kinase activity [51;52]. BCR/ABL induces hematopoietic cell transformation and protects cells from apoptosis [53]. One of the characteristics of CML cells is an early release of myeloid cells from the marrow and their accumulation in the blood and spleen, suggesting that mechanisms that regulate cell retention or migration may be impaired. CML cells show reduced adhesion to stromal cells and extracellular matrix [54;55] that is associated with multiple abnormalities in cytoskeletal function [56]. It is likely that the effects of BCR/ABL on cytoskeletal function, adhesion, increased motility and migration contribute to early exit of transformed cells from bone marrow and subsequent proliferation.

Transformation of hematopoietic cell lines with BCR/ABL increases spontaneous motility; however, it significantly reduces chemotactic response to SDF1. Homing of Ba/F3 cells over-expressing BCR/ABL to spleens that have been ectopically injected with SDF1 was significantly reduced compared to control cells in mice, although there were no differences in the ratio of control and BCR/ABL transformed cells that accumulated in bone marrow, lung or blood. BCR/ABL completely blocks calcium flux and p70 S6 kinase phosphorylation normally induced by SDF1 without changing surface CXCR4 expression in hematopoietic cell lines [37]. Furthermore, BCR/ABL inhibits response to SDF1/CXCR4 by constitutively activating Src-related kinase, Lyn and PI3-kinase [47]. A recent report also indicates that the migratory response of primary CML cells at blast crisis was profoundly reduced, which was associated with reduced CXCR4 expression, while $CD34^+$ cells at chronic phase migrate normally in response to SDF1 [45]. In addition, higher BCR/ABL expression induces marked down regulation of CXCR4, while STI-571 treatment that blocks BCR/ABL activity restores CXCR4 expression in $CD34^+$ cells from patients in blast crisis [45]. The reduction in CXCR4 expression and loss of response to SDF1 mediated by BCR/ABL may allow transformed cells to exit bone marrow, subsequently resulting in infiltration to secondary organs, including blood and spleen. Conversely, SDF1 can increase resistance of acute lymphoblastic leukemia cells expressing BCR/ABL to STI-571 [46], suggesting that SDF1/CXCR4 signaling may augment BCR/ABL signaling that enhances resistance to the therapy.

ITD-Flt3 Alters Chemotaxis and Survival in Response to SDF1/CXCR4 Signaling

SDF1 is the primary chemoattractant for normal HSPC and the majority of AML cells express CXCR4 and migrate in response to SDF1 [21;28], suggesting that the SDF1/CXCR4 axis may be involved in anchoring them within the bone marrow and also in regulating their migration to other organs. Consistent with these findings, CXCR4 is involved in the development of AML in NOD/SCID mice [57]. Clinically, AML cells are found in the systemic bone marrow and often found in extramedullary sites, which itself is an adverse prognostic factor [58;59]. It is conceivable that leukemia cells are originally derived from a single cell in the bone marrow that harbors genetic mutations that subsequently expands, peripheralizes into blood circulation and homes to different organs or infiltrates other bone marrow sites and produces significant hematological complications by impairing production of functional blood cells. In addition, dissemination of AML cells to other marrow sites increases the frequency for AML cells to interact with marrow niches and extracellular matrix (ECM), such as stromal fibronectin, via surface integrins, which is one of the mechanisms responsible for minimal residual disease in AML [34]. Genetic mutations (Internal tandem duplication: ITD) of the Flt3 gene that pathologically auto-activates Flt3 tyrosine kinase activity have been found in ~25-30% of patients with AML and the elevated chemokine receptor CXCR4 expression is associated with poor outcome in patients with acute myeloid leukemia (AML) [22;33], pointing to a role for ITD-Flt3 and CXCR4 in disease progression. Our recent data and that of others demonstrate that the FL/Flt3 axis regulates hematopoietic cell survival/proliferation and migration by activating SDF1/CXCR4 signaling pathways [50;60]. Combination of FL and SDF1 synergistically increases survival and proliferation of normal hematopoietic progenitor cells [15]. Importantly, SDF1 dose-dependently increases growth factor independent proliferation of CFU-GM induced by ITD-Flt3, which was not due to up-regulation of CXCR4 [60]. Moreover, ectopic expression of ITD-Flt3 significantly increases spontaneous cell motility and migration of hematopoietic cells to SDF1 without up-regulating CXCR4 [50]. This is in contrast to BCR/ABL that increases spontaneous motility while reducing migration in response to SDF1 [37;45]. Pretreatment of cells with a selective CXCR4 inhibitor, AMD3100 partially inhibits enhanced migration to SDF1 (unpublished observation). Enhanced chemotactic response to SDF1 is partially mediated through Ras signaling, since dominant negative H-Ras dramatically inhibits spontaneous and SDF-mediated migration of Ba/F3 cells, while constitutively active H-Ras expression in Ba/F3 cells harboring wild-type Flt3 increases chemotactic response to SDF1 as high as ITD-Flt3 [60]. These results suggest that ITD-Flt3 may enhance homing of hematopoietic cells to secondary sites where SDF1 is expressed. Interestingly, ITD-Flt3 mutations also increase cell migration against an SDF1α gradient not just towards SDF1 [50], suggesting that ITD-Flt3 stimulates chemokinesis (random cell migration) and may facilitate peripheralization of malignant hematopoietic cells. Our recent data also demonstrates that ITD-Flt3 reduces adhesion of Ba/F3 cells to VCAM-1 compared to wild-type Flt3 coincident with significant down regulation of $\alpha4$ and $\beta1$ integrin expression [60]. ITD-Flt3 mutations cause extramedullary infiltration of hematopoietic cells in a myeloproliferative disease model in mice [61]. Consistent with this finding, ITD-Flt3 significantly increases accumulation of Ba/F3 cells in spleen shortly after transplantation compared to wild-type Flt3 [60]. Cells circulating in peripheral blood were also elevated by ITD-Flt3 [60]. Although these data indicate that ITD-Flt3 and SDF1/CXCR4 signaling functionally interact to regulate hematopoietic cell survival/transformation and migration, it remains to be determined whether enhanced homing/accumulation to spleen by ITD-Flt3 is mediated by SDF/CXCR4 signaling or other factors such as increased motility or adhesion capability.

Potential Therapeutic Strategy by Blocking SDF1/CXCR4 and Oncoprotein Signaling Interaction for Cancer Treatement

The consequences of over-expression of BCR/ABL, ITD-Flt3 and HER2 on SDF1/CXCR4 function are not identical but share some biological activities (Table).

Table. Oncogenes that regulate SDF/CXCR4 function

Oncoprotein	BCR/ABL	ITD-Flt3	HER2
Disease	**CML**	**AML**	**Breast cancer**
extramedullary infiltration or metastasis in patients	↑circulating blood cells, splenomegaly	↑circulating blood cells , splenomegaly, extramedullary tumor	↑lung metastasis,↑bone marrow metastasis
CXCR4 expression	no change or down regulated	↑in clinical samples expressing ITD-Flt3(↓in Ba/F3 cell lines)	↑
mechanisms responsible for changes in CXCR4 expression	transcriptional		1) up-regulating translation through activating PI3kinase/Akt/ mTOR pathway 2) Inhibiting SDF-induced down regulation
spontaneous motility	↑	↑	
chemotaxis to SDF1	↓	↑	↑
actin polymerization	↑	↑	
signaling in response to SDF	inactivation of S6 kinase &activation of Lyn and PI3kinase	Ras activation is involved in enhanced chemotaxis to SDF	
homing to extramedullary organs or metastasis in mice model	↓spleen homing	↑ spleen homing ↑ circulating cells	↑ lung metastasis
adhesion	↓ adhesion to fibronectin and stromal cells	↓ adhesion to VCAM1	↑ adhesion to endothelial cells
association with poor prognosis and co-expression with CXCR4	↓CXCR4 at blast crisis	+	+
effect of SDF on cell survival	↑resistance to STI-571	↑ growth factor independent survival of primary mouse HSPC	↑ survival of breast cancer cells
transactivation by SDF			+

In contrast to ITD-Flt3 and HER2, BCR/ABL reduces in vitro migration, signaling and in vivo homing to spleen in response to SDF1. Although reduced CXCR4 expression and migration to SDF1 are not observed during the chronic phase of CML, they are associated with blast crisis and are restored by treatment with the BCR/ABL kinase inhibitor, STI-571 [37;45]. These findings suggest that functional alteration of CXCR4 by BCR/ABL may lead to disease progression of CML. BCR/ABL positive lympoblastic cells become resistance to STI-571 when cultured with SDF1 or in the presence of stromal cells that produce SDF1 [46]. However, AMD3100, the selective antagonist for CXCR4, reduces the enhanced resistance to STI-571 induced by stromal cells, indicating that blocking CXCR4 signaling can reduce aberrant signaling generated by BCR/ABL. These results suggest that BCR/ABL and CXCR4 signaling modulate each other and that blocking these pathways may enhance sensitivity to therapeutic modalities, although their consequence on cell proliferation, migration and invasion in CML cells remains to be determined.

In contrast to HER2, ITD-Flt3 does not up-regulate CXCR4 expression; however the biological effect produced by ITD-Flt3 is similar to HER2 with respect to response to SDF1. Both ITD-Flt3 and HER2 increase cell migration to SDF1 and enhance homing or metastasis to secondary organs (spleen and lung by ITD-Flt3 and HER2, respectively) [60, 49]. These results also indicate that ITD-Flt3 and HER2 signaling activate SDF1/CXCR4 signaling, although it is not clear whether the enhanced homing of hematopoietic cells expressing ITD-Flt3 to spleen is completely dependent on SDF1/CXCR4 signaling. Conversely, SDF1 increases growth factor independent survival of primary mouse hematopoietic progenitor cells induced by ITD-Flt3 [60]. SDF1 also transactivates HER2 in breast cancer cell lines while AMD3100 inhibits this effect [48], indicating that SDF1/CXCR4 signaling can also activate ITD-Flt3 and HER2 signaling. Blocking SDF1/CXCR4 inhibits breast cancer metastasis and Herceptin, an inhibitor of HER2, reduces their migration to SDF1. The fact that SDF1 is a survival factor for breast cancer cells [43] further suggests that blocking both CXCR4 plus HER2 may provide an additional therapeutic strategy for breast cancers.

Since SDF1 significantly enhances ITD-Flt3 mediated growth factor independent proliferation of CFU-GM and SDF1 is produced by stromal cells and osteoblastic cells [3;62], interactions between AML cells harboring ITD-Flt3 and these cells may increase their survival as a consequence of ITD-Flt3 and CXCR4 signaling cross talk. ITD-Flt3 modulates migration in response to SDF1 and trafficking of hematopoietic cells in vivo. These results suggest that: 1) interactions between ITD-Flt3 and SDF1/CXCR4 signaling may be responsible for aberrant hematopoiesis in patients with AML by enhancing survival/transformation or dissemination of leukemia cells; 2) interactions between the bone marrow microenvironment and cells expressing ITD-Flt3 may synergistically increase their survival through SDF1/CXCR4 signaling, which may lead to resistance to apoptosis; and 3) manipulation of ITD-Flt3 and SDF1/CXCR4 signaling interactions may be clinically beneficial for the treatment of hematopoietic malignancies in which both ITD-Flt3 and CXCR4 are expressed.

Conclusion

The emerging role of SDF1/CXCR4 function in tumor cell survival, invasion, metastasis and poor prognosis of patients with various tumors makes antagonizing the SDF1/CXCR4 pathway as a candidate strategy for anti-cancer therapy. Antagonizing SDF1/CXCR4 signaling pathways has proven to be useful to suppress growth, invasion or metastasis of several cancer cells [49;63] or hematological malignancies [29-31] in experimental models. The functional cross talk between BCR/ABL, ITD-Flt3, HER2 and SDF1/CXCR4 signaling appears to modulate survival, dissemination or metastasis of tumor cells. Concomitant expression of CXCR4 and these oncoproteins is associated with poor patient prognosis. Simultaneous inhibition of SDF1/CXCR4 and ITD-Flt3, BCR/ABL or HER2 may provide a more effective treatment modality in patients with tumors expressing these molecules. Identification of mechanism responsible for signaling interactions between SDF/CXCR4 and these oncogenes may define additional therapeutic targets. However, consequences of prolonged CXCR4 disruption in normal cells has not been clearly determined and understanding the role of SDF1/CXCR4 function in normal versus malignant cells will be important in identifying strategies that maximally disrupt SDF1/CXCR4 in cancer cells with minimal effect on normal tissues.

[1] References

[2] Tashiro K, Tada H, Heilker R, Shirozu M, Nakano T, Honjo T. Signal sequence trap: a cloning strategy for secreted proteins and type I membrane proteins. *Science* 1993; 261:600-3.

[3] Nagasawa T, Kikutani H, Kishimoto T. Molecular cloning and structure of a pre-B-cell growth-stimulating factor. *Proc Natl Acad Sci U S A* 1994; 91:2305-9.

[4] Jung Y, Wang J, Schneider A et al. Regulation of SDF-1 (CXCL12) production by osteoblasts; a possible mechanism for stem cell homing. *Bone* 2006; 38:497-508.

[5] Semerad CL, Christopher MJ, Liu F et al. G-CSF potently inhibits osteoblast activity and CXCL12 mRNA expression in the bone marrow. *Blood* 2005; 106:3020-7.

[6] Tachibana K, Hirota S, Iizasa H et al. The chemokine receptor CXCR4 is essential for vascularization of the gastrointestinal tract. *Nature* 1998; 393:591-4.

[7] Zou YR, Kottmann AH, Kuroda M, Taniuchi I, Littman DR. Function of the chemokine receptor CXCR4 in haematopoiesis and in cerebellar development. *Nature* 1998; 393:595-9.

[8] Feng Y, Broder CC, Kennedy PE, Berger EA. HIV-1 entry cofactor: functional cDNA cloning of a seven-transmembrane, G protein-coupled receptor. *Science* 1996; 272: 872-7.

[9] Peled A, Petit I, Kollet O et al. Dependence of human stem cell engraftment and repopulation of NOD/SCID mice on CXCR4. *Science* 1999; 283:845-8.

[10] Levesque JP, Hendy J, Takamatsu Y, Simmons PJ, Bendall LJ. Disruption of the CXCR4/CXCL12 chemotactic interaction during hematopoietic stem cell mobilization induced by G-CSF or cyclophosphamide. *J Clin Invest* 2003; 111:187-96.

[11] Kim CH, Broxmeyer HE. In vitro behavior of hematopoietic progenitor cells under the influence of chemoattractants: stromal cell-derived factor-1, steel factor, and the bone marrow environment. *Blood* 1998; 91:100-10.

[12] Christopherson KW 2nd, Cooper S, Broxmeyer HE. Cell surface peptidase CD26/DPPIV mediates G-CSF mobilization of mouse progenitor cells. *Blood* 2003; 101:4680-6.

[13] Liles WC, Broxmeyer HE, Rodger E et al. Mobilization of hematopoietic progenitor cells in healthy volunteers by AMD3100, a CXCR4 antagonist. *Blood* 2003; 102: 2728-30.

[14] Peled A, Kollet O, Ponomaryov T et al. The chemokine SDF-1 activates the integrins LFA-1, VLA-4, and VLA-5 on immature human CD34(+) cells: role in transendothelial/stromal migration and engraftment of NOD/SCID mice. *Blood* 2000; 95:3289-96.

[15] Broxmeyer HE, Kohli L, Kim CH et al. Stromal cell-derived factor-1/CXCL12 directly enhances survival/antiapoptosis of myeloid progenitor cells through CXCR4 and G(alpha)*i* proteins and enhances engraftment of competitive, repopulating stem cells. *J Leukoc Biol* 2003; 73:630-8.

[16] Lee Y, Gotoh A, Kwon HJ et al. Enhancement of intracellular signaling associated with hematopoietic progenitor cell survival in response to SDF-1/CXCL12 in synergy with other cytokines. *Blood* 2002; 99:4307-17

[17] Ma Q, Jones D, Springer TA. The chemokine receptor CXCR4 is required for the retention of B lineage and granulocytic precursors within the bone marrow microenvironment. *Immunity* 1999; 10:463-71.

[18] Ma Q, Jones D, Borghesani PR et al. Impaired B-lymphopoiesis, myelopoiesis, and derailed cerebellar neuron migration in CXCR4- and SDF-1-deficient mice. *Proc Natl Acad Sci U S A* 1998; 95:9448-53.

[19] Nagasawa T, Hirota S, Tachibana K et al. Defects of B-cell lymphopoiesis and bone-marrow myelopoiesis in mice lacking the CXC chemokine PBSF/SDF-1. *Nature* 1996; 382:635-8.

[20] Burger JA, Burger M, Kipps TJ. Chronic lymphocytic leukemia B cells express functional CXCR4 chemokine receptors that mediate spontaneous migration beneath bone marrow stromal cells. *Blood* 1999; 94:3658-67.

[21] Bradstock KF, Makrynikola V, Bianchi A, Shen W, Hewson J, Gottlieb DJ. Effects of the chemokine stromal cell-derived factor-1 on the migration and localization of precursor-B acute lymphoblastic leukemia cells within bone marrow stromal layers. *Leukemia* 2000; 14:882-8.

[22] Mohle R, Schittenhelm M, Failenschmid C et al. Functional response of leukaemic blasts to stromal cell-derived factor-1 correlates with preferential expression of the chemokine receptor CXCR4 in acute myelomonocytic and lymphoblastic leukaemia. *Br J Haematol* 2000; 110:563-72.

[23] Rombouts EJ, Pavic B, Lowenberg B, Ploemacher RE. Relation between CXCR-4 expression, Flt3 mutations, and unfavorable prognosis of adult acute myeloid leukemia. *Blood* 2004; 104:550-7.

[24] Voermans C, van Heese WP, de Jong I, Gerritsen WR, van Der Schoot CE. Migratory behavior of leukemic cells from acute myeloid leukemia patients. *Leukemia* 2002; 16:650-7.

[25] Trentin L, Cabrelle A, Facco M et al. Homeostatic chemokines drive migration of malignant B cells in patients with non-Hodgkin lymphomas. *Blood* 2004; 104:502-8.

[26] Weng AP, Shahsafaei A, Dorfman DM. CXCR4/CD184 immunoreactivity in T-cell non-Hodgkin lymphomas with an overall Th1- Th2+ immunophenotype. *Am J Clin Pathol* 2003; 119:424-30.

[27] Sanz-Rodriguez F, Hidalgo A, Teixido J. Chemokine stromal cell-derived factor-1α modulates VLA-4 integrin-mediated multiple myeloma cell adhesion to CS-1/fibronectin and VCAM-1. *Blood* 2001; 97:346-51.

[28] Burger JA, Kipps TJ. CXCR4: a key receptor in the crosstalk between tumor cells and their microenvironment. *Blood* 2006; 107:1761-7.

[29] Mohle R, Bautz F, Rafii S, Moore MAS, Brugger W, Kanz L. The chemokine receptor CXCR-4 is expressed on CD34+ hematopoietic progenitors and leukemic cells and mediates transendothelial migration induced by stromal cell-derived factor-1. *Blood* 1998; 91:4523-30.

[30] Juarez J, Bradstock KF, Gottlieb DJ, Bendall LJ. Effects of inhibitors of the chemokine receptor CXCR4 on acute lymphoblastic leukemia cells in vitro. *Leukemia* 2003; 17:1294-300.

[31] Burger M, Hartmann T, Krome M et al. Small peptide inhibitors of the CXCR4 chemokine receptor (CD184) antagonize the activation, migration, and antiapoptotic responses of CXCL12 in chronic lymphocytic leukemia B cells. *Blood* 2005; 106: 1824-30.

[32] Bertolini F, Dell'Agnola C, Mancuso P et al. CXCR4 neutralization, a novel therapeutic approach for non-Hodgkin's lymphoma. Cancer Res 2002; 62:3106-12.

[33] Crazzolara R, Kreczy A, Mann G et al. High expression of the chemokine receptor CXCR4 predicts extramedullary organ infiltration in childhood acute lymphoblastic leukaemia. *Br J Haematol* 2001; 115:545-53.

[34] Spoo AC, Lubbert M, Wierd WG, Burger JA. CXCR4 is a prognostic marker in Acute Myelogenous Leukemia. Blood 2007; 109: 786-791.

[35] Matsunaga T, Takemoto N, Sato T et al. Interaction between leukemic-cell VLA-4 and stromal fibronectin is a decisive factor for minimal residual disease of acute myelogenous leukemia. *Nature Medicine* 2003; 9:1158-65.

[36] Muller A, Homey B, Soto H et al. Involvement of chemokine receptors in breast cancer metastasis. *Nature* 2001; 410:50-6.

[37] Burger M, Glodek A, Hartmann T et al. Functional expression of CXCR4 (CD184) on small-cell lung cancer cells mediates migration, integrin activation, and adhesion to stromal cells. *Oncogene* 2003; 22:8093-101.

[38] Salgia R, Quackenbush E, Lin J et al. The BCR/ABL oncogene alters the chemotactic response to stromal-derived factor-1α. *Blood* 1999; 94:4233-46.

[39] Wang J, Wang J, Sun Y et al. Diverse signaling pathways through the SDF-1/CXCR4 chemokine axis in prostate cancer cell lines leads to altered patterns of cytokine secretion and angiogenesis. *Cell Signal* 2005; 17:1578-92.

[40] Scotton CJ, Wilson JL, Scott K et al. Multiple actions of the chemokine CXCL12 on epithelial tumor cells in human ovarian cancer. *Cancer* Res 2002; 62:5930-8.

[41] Geminder H, Sagi-Assif O, Goldberg L et al. A possible role for CXCR4 and its ligand, the CXC chemokine stromal cell-derived factor-1, in the development of bone marrow metastases in neuroblastoma. *J Immunol* 2001; 167:4747-57.

[42] Scala S, Ottaiano A, Ascierto PA et al. Expression of CXCR4 predicts poor prognosis in patients with malignant melanoma. *Clin Cancer* Res 2005; 11:1835-41.

[43] Zhou Y, Larsen PH, Hao C, Yong VW. CXCR4 is a major chemokine receptor on glioma cells and mediates their survival. *J Biol Chem* 2002; 277:49481-7.

[44] Orimo A, Gupta PB, Sgroi DC et al. Stromal fibroblasts present in invasive human breast carcinomas promote tumor growth and angiogenesis through elevated SDF-1/CXCL12 secretion. *Cell* 2005; 121:335-48.
[45] Burger JA, Tsukada N, Burger M, Zvaifler NJ, Dell'Aquila M, Kipps TJ. Blood-derived nurse-like cells protect chronic lymphocytic leukemia B cells from spontaneous apoptosis through stromal cell-derived factor-1. *Blood* 2000; 96:2655-63.
[46] Geay JF, Buet D, Zhang Y et al. p210BCR-ABL inhibits SDF-1 chemotactic response via alteration of CXCR4 signaling and down-regulation of CXCR4 expression. *Cancer Res* 2005; 1:7-2676.
[47] Mishra S, Zhang B, Cunnick JM, Heisterkamp N, Groffen J. Resistance to imatinib of bcr/abl p190 lymphoblastic leukemia cells. *Cancer Res* 2006; 66:5387-93.
[48] Ptasznik A, Urbanowska E, Chinta S et al. Crosstalk between BCR/ABL oncoprotein and CXCR4 signaling through a Src family kinase in human leukemia cells. *J Exp Med* 2002; 196:667-78.
[49] Cabioglu N, Summy J, Miller C et al. CXCL-12/stromal cell-derived factor-1α transactivates HER2-neu in breast cancer cells by a novel pathway involving Src kinase activation. *Cancer Res* 2006; 65:6493-7.
[50] Li YM, Pan Y, Wei Y et al. Upregulation of CXCR4 is essential for HER2-mediated tumor metastasis. *Cancer Cell* 2004; 6:459-69.
[51] Fukuda S, Broxmeyer HE, Pelus LM. Flt3 ligand and the Flt3 receptor regulate hematopoietic cell migration by modulating the SDF-1α(CXCL12)/CXCR4 axis. *Blood* 2005; 105:3117-26.
[52] Sawyers CL. Chronic myeloid leukemia. *N Engl J Med* 1999; 340:1330-40.
[53] Kurzrock R, Gutterman JU, Talpaz M. The molecular genetics of Philadelphia chromosome-positive leukemias. *N Engl J Med* 1988; 319:990-8.
[54] Amarante-Mendes GP, Naekyung Kim C, Liu L et al. Bcr-Abl exerts its antiapoptotic effect against diverse apoptotic stimuli through blockage of mitochondrial release of cytochrome C and activation of caspase-3. *Blood* 1998; 91:1700-5.
[55] Gordon MY, Dowding CR, Riley GP, Goldman JM, Greaves MF. Altered adhesive interactions with marrow stroma of haematopoietic progenitor cells in chronic myeloid leukaemia. *Nature* 1987; 328:342-4.
[56] Verfaillie CM, McCarthy JB, McGlave PB. Mechanisms underlying abnormal trafficking of malignant progenitors in chronic myelogenous leukemia. Decreased adhesion to stroma and fibronectin but increased adhesion to the basement membrane components laminin and collagen type IV. *J Clin Invest* 1992; 90:1232-41.
[57] Salgia R, Li JL, Ewaniuk DS et al. BCR/ABL induces multiple abnormalities of cytoskeletal function. *J Clin Invest* 1997; 100:46-57.
[58] Tavor S, Petit I, Porozov S et al. CXCR4 regulates migration and development of human acute myelogenous leukemia stem cells in transplanted NOD/SCID mice. *Cancer Res* 2004; 64:2817-24.
[59] Reinhardt D, Pekrun A, Lakomek M, Zimmermann M, Ritter J, Creutzig U. Primary myelosarcomas are associated with a high rate of relapse: report on 34 children from the acute myeloid leukaemia-Berlin-Frankfurt-Munster studies. *Br J Haematol* 2000; 110:863-6.

[60] Byrd JC, Edenfield WJ, Shields DJ, Dawson NA. Extramedullary myeloid cell tumors in acute nonlymphocytic leukemia: a clinical review. *J Clin Oncol* 1995; 13:1800-16..

[61] Fukuda S, Pelus LM. Internal Tandem Duplication of Flt3 (ITD-Flt3) modulates chemotaxis and survival of hematopoietic cells by SDF1α but negatively regulates marrow homing in vivo. *Exp Hematol* 2006; 34:1041-51.

[62] Kelly LM, Liu Q, Kutok JL, Williams IR, Boulton CL, Gilliland DG. FLT3 internal tandem duplication mutations associated with human acute myeloid leukemias induce myeloproliferative disease in a murine bone marrow transplant model. *Blood* 2002; 99:310-8.

[63] Ponomaryov T, Peled A, Petit I et al. Induction of the chemokine stromal-derived factor-1 following DNA damage improves human stem cell function. *J Clin Invest* 2000; 106:1331-9.

[64] Chen Y, Stamatoyannopoulos G, Song CZ. Down-regulation of CXCR4 by inducible small interfering RNA inhibits breast cancer cell invasion in vitro. *Cancer* Res 2003; 63:4801-4.

In: Chemokine Research Trends
Editor: Leah R. Grinwald, pp. 159-169
ISBN: 978-60021-588-9

Chapter 11

ROLES OF PLATELET-DERIVED CHEMOKINES IN VARIOUS CLINICAL SETTINGS

***Shodsaku Nomura**[*1]* ***and Norihito Inami**[2]*

[1] Division of Hematology, Kishiwada City Hospital, Kishiwada, Osaka, Japan
[2] Second Department of Internal Medicine, Kansai Medical University, Moriguchi, Osaka, Japan

ABSTRACT

Chemokines are a superfamily of chemotactic cytokines. Recently, platelets have been shown to contain members of the chemokine family, both of CXC class, such as interleukin (IL)-8 and epithelial-neutrophil activating peptide (ENA)-78, and of the C-C class, such as regulated on activation normally T-cell expressed and secreted (RANTES). RANTES is a potent chemoattractant of memory T lymphocytes, monocytes, eosinophils and basophils. And ENA-78 is chemokine known to be chemoattractant for neutrophils. We measured and compared levels of platelet activation markers, platelet-derived chemokines, and soluble factors in patients with various diseases and clinical settings. RANTES differed in the healthy control and hypertensive patients, particularly with type 2 diabetes. In addition, this chemokine was higher in the patients with acute myocardial infarction than in patients with stable pectoris angina. ENA-78 concentration was found in Helicobacter pylori-positive patients with immune thrombocytopenic purpura. Furthermore, ENA-78 and RANTES exhibited a significant elevation after stem cell transplantation. Thus, platelet-derived chemokines demonstrate the role of platelets as inflammatory cells in various clinical settings.

INTRODUCTION

Chemokines are a superfamily of chemotactic cytokines. They activate and direct the migration of leucocytes [1-3]. Monocyte chemotactic peptide-1 (MCP-1) is a C-C chemokine for monocytes that can amplify the inflammatory response by recruiting additional peripheral blood monocytes across the vascular endothelium to an inflammatory site [4,5]. Regulated on

* E-mail address: shosaku-n@mbp.ocn.ne.jp, Tel: + 81 724 45 1000, Fax: + 81 725 32 1113, Division of Hematology, Kishiwada City Hospital1001, Gakuhara-cho, Kishiwada Osaka 596-8501, JAPAN

activation normally T-cell expressed and secreted (RANTES), another C-C chemokine, is a potent chemoattractant for memory T lymphocytes, monocytes, eosinophils and basophils [6]. IL-8, epithelial-neutrophil activating peptide (ENA)-78, and growth-regulated oncogene (GRO) α are CXC chemokines known to be chemoattractants for neutrophils [6,7]. Platelets have been shown to contain members of the chemokine family, both those of the CXC class, such as IL-8 and ENA-78 [8], and those of the C-C class, such as RANTES [9]. This report demonstrates the role of platelets as inflammatory cells. We present data comparing the levels of platelet activation markers, platelet-derived chemokines, and soluble factors in patients with various diseases and in various clinical settings.

Type 2 Diabetes Mellitus

The levels of cell adhesion molecules have been reported to be increased in both type 1 and type 2 diabetes [10], and have been implicated in the development of microvascular complications. In particular, E-selectin is upregulated at sites of atherosclerotic lesions [11], and serum levels of soluble E-selectin (sE-selectin) are elevated in patients with diabetes [10]. Monocytes and macrophages may play an important role in glomerular disease [12] and it has been reported that MCP-1 might contribute to glomerular monocyte infiltration in diabetic nephropathy [13]. However, no in vivo findings exist concerning RANTES levels in diabetic patients. Therefore, we measured and compared the levels of platelet activation markers, chemokines, and soluble selectins, in order to develop a better understanding of their potential contribution to diabetic vascular complications [14]. There were no differences in IL-8 levels between the control and diabetic groups; however, the levels of MCP-1, RANTES, and soluble factors were higher in diabetic patients compared with controls (MCP-1: 345±19 vs. 234±17 pg/ml, $p < 0.05$; RANTES: 112±25 vs. 41±5 ng/ml, $p < 0.01$). The IL-8 level did not significantly correlate with that of platelet-P-selectin, but the MCP-1 and RANTES levels were both positively correlated with the level of platelet-P-selectin ($p < 0.05$ and $p < 0.001$, respectively).

We also evaluated the plasma concentrations of chemokines and soluble selectins in diabetic patients; however, it is unclear whether the serum levels of P- and E-selectin reflect the tissue levels of these selectins. In our study, the levels of the two C-C chemokines MCP-1 and RANTES were increased in diabetic patients compared with healthy subjects. Enhanced levels of C-C chemokines have been found in inflammatory and immune-mediated diseases [15-17]. Thus, we believe that our findings support the notion that immunological and inflammatory processes are important features of diabetic complications. MCP-1 and RANTES are produced by a variety of leukocytes; RANTES is also produced by platelets, whereas MCP-1 is also produced by endothelial cells and fibroblasts [9,18]. In our study, both RANTES and MCP-1 levels were positively correlated with the level of platelet-P-selectin. Activated platelets have been found to stimulate MCP-1 production in monocytes through enhanced RANTES secretion and direct platelet-monocyte contact mediated by P-selectin expression on the platelet surface [19]. In addition, levels of MCP-1 and RANTES were different in hypertensive patients with and without type 2 diabetes mellitus [20-22].

Coronary Artery Diseases

Evidence from many clinical and epidemiological studies suggests that the evolution from unstable disease to acute thrombosis is linked to several hemostatic and hemorheologic factors [23]. In addition, certain clinical procedures, such as percutaneous coronary intervention (PCI), are linked to platelet activation, leading to acute ischemic events in treated patients [24]. An increased leukocyte count in acute myocardial infarction (AMI) patients is an important risk factor for subsequent adverse cardiac events [25]. The procoagulant activity of leukocytes may contribute to the risk of thrombotic events after AMI [26]. Surface expression of tissue factor on activated monocytes might play a central role in cellular procoagulant responses. A number of cytokines can induce leukocyte procoagulant activity [27]. RANTES may also activate specific effector cell populations of leukocytes [9].

Markers of platelet activation have been detected in the peripheral blood of patients experiencing an acute ischemic event after PCI and also after direct PCI for AMI [28]. Platelet-derived microparticles (PDMPs) are produced by platelet activation or by physical stimulation under various conditions [29]. Occasional mention of the potential role of PDMPs is found in some reports on acute coronary syndrome [28,30,31]. Therefore, we compared the levels of PDMPs and RANTES in order to evaluate changes in platelet and chemokine function in patients undergoing PCI.

The levels of PDMP and RANTES were significantly higher in patients with acute coronary syndrome (ACS) than those in the control groups (PDMP: 20.1 ± 2.9 vs 80.4 ± 7.3 U/ml, $p < 0.001$; RANTES: 18.6 ± 3.7 vs 52.1 ± 4.6 ng/ml, $p < 0.01$); however, the levels did not differ between the control groups and patients with stable angina. PDMP levels decreased significantly after PCI treatment. In particular, in patients with AMI, a time-dependent decrease in PDMP level was observed. RANTES levels also decreased significantly after PCI ($p < 0.05$), but there was no difference between the two groups. In addition, the correlations between PDMP and either RANTES or sCD40L, in measured samples before and after PCI, were significant (PDMP vs. RANTES, $r=0.598$, $p < 0.01$; PDMP vs. sCD40L, $r=0.672$, $p < 0.001$). Thus, PDMP and RANTES levels were shown to be significantly elevated in patients with ACS, and the levels were significantly decreased after PCI. Increased levels of C-C chemokines have been found in association with inflammatory and immune-mediated diseases [15-17]. Thus, our findings support the notion that immunologic and inflammatory processes are important factors in ACS. RANTES is produced by a variety of leukocytes and platelets [9,18]. The significant positive correlation we observed between RANTES and PDMP levels in this study indicates that RANTES originates in patients with ACS in relation to both platelet and leukocyte activation. Activated platelets have been found to stimulate MCP-1 production through enhanced RANTES secretion and direct platelet-monocyte contact mediated by P-selectin on the platelet surface [19].

PDMP and RANTES levels were also significantly decreased after PCI in patients with ACS. PCI and antiplatelet therapy might suppress the generation of vascular complications in which T-lymphocytes and monocytes participate, as RANTES levels are decreased by a combination of PCI and anti-platelet therapy. Thus, monitoring and measuring RANTES and PDMP levels might be useful for evaluating clinical condition after PCI in patients with ACS.

IMMUNE THROMBOCYTOPENIC PURPURA

Immune thrombocytopenic purpura (ITP) is an autoimmune disease caused by circulating antibodies that react with platelet membranes [32]. Emilia et al [33] reported a high incidence of Helicobacter pylori (HP) infection in patients with ITP as well as a significant increase in platelet count after bacterium eradication. Although the pathogenetic mechanisms of HP-induced thrombocytopenia remain unknown, HP causes an immunological response resulting in the production of high concentrations of pro-inflammatory cytokines and chemokines in non-ITP patients [34]. On the contrary, we reported on the significance of chemokines in patients with ITP [35]. MCP-1 and RANTES were significantly increased in ITP. This elevation was thought to be related to the CD40/CD40 ligand system. We thus investigated the concentrations of different chemokines before and after HP eradication in ITP patients.

Serum samples from 42 patients with chronic ITP were studied. The urea breath test was used to determine whether or not patients were infected with HP, and if found to be infected, treatment to eradicate HP included amoxicillin (1000 mg twice daily), clarithromycin (400 mg twice daily), and Omeprazole (20 mg twice daily) for 1 week. The prednisolone treatment was not changed during the eradication therapy, which was evaluated at 8 week following treatment. We measured human 3 C-C chemokines (MCP-1, RANTES, and Eotaxin) and 3 CXC chemokines (IL-8, ENA-78, and stromal cell-derived factor ; SDF-1) because HP influences both C-C and CXC chemokine levels [36,37].

HP infection was found in 28 of 42 chronic ITP patients (66.7%). There were no significant differences in concentrations of IL-8, SDF-1 and Eotaxin between HP-positive and HP-negative ITP patients. However, significant increases in MCP-1, RANTES, and ENA-78 concentrations were found in HP-positive ITP patients. Chemokines remained elevated although the platelet counts increased; however, RANTES and ENA-78 concentrations exhibited non-significant decrease in HP eradication responders.

Increased production of several cytokine/chemokine factors, by immunocompetent cells such as lymphocytes or monocytes suggest they participate in some immunological events [38], and elevated concentrations of different chemokines have been seen in some immune-mediated diseases [15,17]. Based on these findings the immunological change in cytokine/chemokine levels may possibly affect the thrombocytopenic mechanism produced by HP infection. However, some researchers do not support the participation of HP in ITP [39,40]. In the present study, we did not find any positive chemokine production effect from the HP eradication therapy. Further research is necessary to understand the mechanism of thrombocytopenia in HP-positive ITP patients.

STEM CELL TRANSPLANTATION

Stem cell transplantation (SCT) is being used for hematopoietic reconstitution following high-dose chemotherapy for malignancy [41-42]. Many cytokines are known to control this hematopoiesis positively or negatively in vivo [43]. A high serum concentration of tumor necrosis factor alpha (TNFα) precedes the onset of major transplant-related complications [44]. Elevated serum levels of interleukin-6 (IL-6) and macrophage-colony stimulating factor (M-CSF) have been reported in patients developing major organ toxicity or graft-versus-host disease (GVHD)[45-46]. Thus, some patients seem to have an imbalance of the immune response after SCT, and cytokines are known to regulate this response [43,47]. We measured and compared levels of platelet activation markers, chemokines, and soluble factors in patients undergoing SCT [48].

IL-8 and GROα exhibited a significant elevation in the early phase (1 or 2 weeks) after SCT, particularly after auto SCT (Figure 1). Furthermore, these levels significantly and positively correlated with the change in G-CSF. IL-8 and GROα belong to the CXC chemokine family and are also potent neutrophil chemoattractants. The relationship between IL-8 and GROα and G-CSF suggests that the roles of IL-8 and GROα may be related to hematopoietic conditions in which G-CSF is involved.

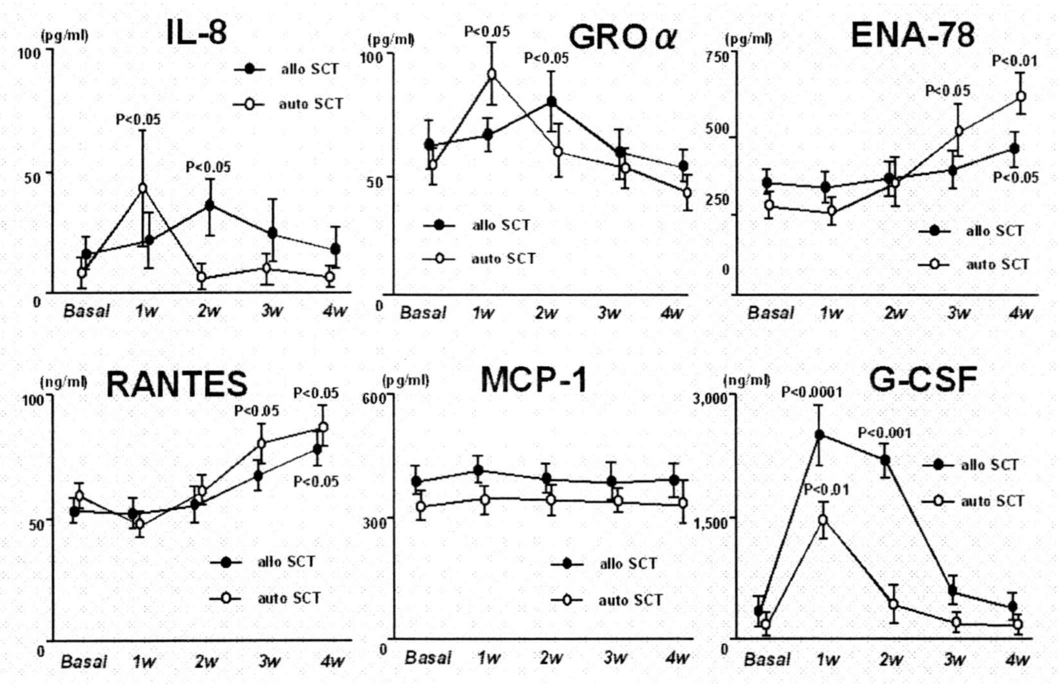

Figure 1. Serum levels of interleukin-8 (IL-8), growth-regulated oncogene alpha (GROα), epithelial neutrophil-activating peptide 78 (ENA-78), monocyte chemoattractant protein-1 (MCP-1), "regulated on activation normally T-cell expressed and secreted" (RANTES), and granulocyte-cplony stimulating factor (G-CSF) in patients undergoing allo- and auto-SCT. Blood samples were obtained on days 0 (basal), 7 (1 w), 14 (2 w), 21 (3 w) and 28 (4 w) after transplantation.: allo-SCT, ○ : auto-SCT. Data represent the mesns and S.E. Student-*T* test was used for statistical comparisons.

In contrast, ENA-78 exhibited a significant elevation in the later phase (3 or 4 weeks) after SCT (Figure 1). In addition, its level correlated negatively with the change in G-CSF. This result suggests that ENA-78 plays a role in hematopoietic conditions in which G-CSF is not involved. ENA-78 is generated from various sources, and one of which is platelets [49-51]. Therefore, platelet activation is a possible cause of elevation of ENA-78. In our study, sCD40L and PDMP increased significantly after both auto and allo SCT. In addition, ENA-78 positively correlated with the level of PDMP. These results suggest different roles for various neutrophil attractant chemokines. When G-CSF is given by injection such as in the early phase after SCT, IL-8 and GROα could be supported by the effects of G-CSF. And when the

G-CSF level returns to baseline in peripheral blood in the later phase after SCT, ENA-78 could cause stabilization of the neutrophil level, which is dependent on platelet activation.

Another important platelet-related chemokine is RANTES [18]. In our study, the increase of RANTES seems to be related to platelet activation, since RANTES was in the dynamic phase similar to sCD40L and PDMP. This increase in RANTES may be involved in the mobilization of T cells after allo SCT. We also measured some cytokines and soluble factors in the present study. It has previously been reported that some cytokines and soluble factors are useful for the diagnosis of GVHD after allo SCT. Proinflammatory cytokines including IFNγ, IL-6, and TNFα are important mediators and regulators of GVHD [52]. sIL-2R appears to be a convenient marker for the detection of acute GVHD [53]. In our study, the involvement of platelet activation in the elevation of IL-6, TNFα, and sIL-2R after allo SCT appeared to be important, since RANTES exhibited the same changes as these cytokines and soluble factors. PDMP also exhibited changes similar to RANTES. This suggests the possibility that RANTES, a platelet activation-dependent chemokine, plays a role in GVHD after allo SCT. Although the molecular pathogenesis of GVHD remains to be uncovered, there is general agreement that infiltrating T lymphocytes play a central role [54-55]. In addition, chemokines such as MCP-1 and RANTES might play important roles in the early pathogenesis of GVHD [56]. Murai et al [55] reported that CCR5-expressing $CD8^+$ T cells participate in the pathogenesis of liver injury in GVHD. CCR5 is preferentially expressed by Th1-type cells, and is the receptor for MIP and RANTES [56,57], suggesting that $CCR5^+$ T-cells accumulate selectively in response to RANTES. In the present study, RANTES exhibited changes similar to IL-6, TNFα, and sIL-2R, which are GVHD markers, as in previous reports [58-59]. Thus, when platelets are activated after SCT, it may cause generation of platelet-derived RANTES. In this manner, RANTES generation after allo SCT seems to relate to GVHD.

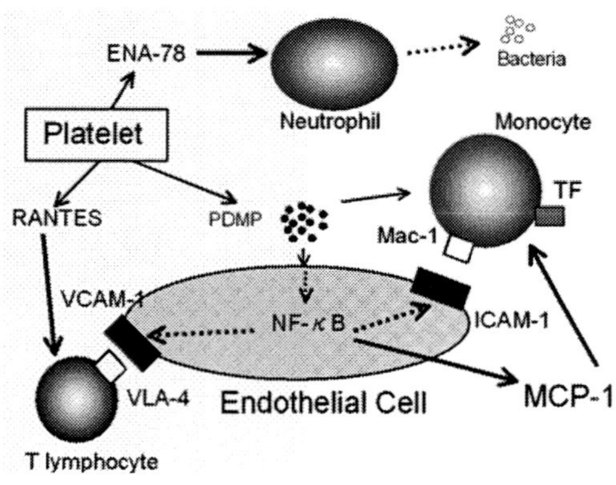

Figure 2. Role of platelet-derived chemokines.The increase of chemokine levels caused by platelet activation. Platelet activation is a feature of many inflammatory diseases. Platelet microparticles (PDMP) release from activated platelets can induce monocyte/endothelial cell activation, tissue factor (TF) expression and MCP-1 generation from activated monocyte and/or endothelial cell. In addition, ENA-78 can induce neutrophil chemotaxis.

CONCLUSION

We have presented data comparing the levels of platelet activation markers, platelet-derived chemokines, and soluble factors in patients with various diseases and in various clinical settings. Our findings suggest that platelet-derived chemokines play an important role in various immune responses and vascular complications (Figure 2). In addition, we show a role for platelet-derived chemokines in novel clinical settings [60-62]. Further examination will be necessary to establish the clinical significane of platelet-derived chemokines.

REFERENCES

[1] Luster AD. Chemokines chemotactic cytokines that mediate inflammation. *N Engl J Med* 1998; 338: 436-445.

[2] Gerard C, Rollins BJ. Chemokines and disease. *Nat Immunol* 2001; 2: 108-115.

[3] Wong MM, Fish EN. Chemokines: attractive mediatirs of the immune response. *Semin Immunol* 2003; 15: 5-14.

[4] Leonard EJ, Yoshimura T. Human monocyte chemoattractant protein-1 (MCP-1). *Immunol Today* 1990; 11: 97-101.

[5] Colotta F, Borre A, Wang JM, Tattanelli M, Maddalena F, Polentarutti N, Peri G, Mantovani A. Expression of a monocyte chemotactic cytokine by human mononuclear phagocytes. *J Immunol* 1992; 148: 760-765.

[6] Baggiolini M, Dewald B, Moster B. Interleukin-8 and related chemotactic cytokines – CXC and CC chemokine. *Adv Immunol* 1994; 55: 97-179.

[7] Ahuja SK, Murphy PM. The CXC chemokines growth-regulated oncogene (GRO) α, GRO β, GRO γ, neutrophil-activating peptide-2, and epithelial cell-derived neutrophil-activating peptide-78 are potent agonists for the type B, but not the type A, human interleukin-8 receptor. *J Biol Chem* 1996; 271: 20545- 20550.

[8] Power CA, Clemetson JM, Clemetson KJ, Wells TN. Chemokine and chemokine receptor mRNA expression in human platelets. *Cytokine* 1995; 7: 479-482.

[9] Kameyoshi Y, Dorschner A, Mallet AI, Christphers E, Schroder JM. Cytokine RANTES released by thrombin-stimulated platelets is a potent attractant for human eosinophils. *J Exp Med* 1992; 176: 587-592.

[10] Cominacini L, Pasini AF, Garbin U, Davoli A, De Santis A, Campagnola M, Rigoni A, Zenti MG, Moeghetti P, Lo Cascio V. Elevated levels of soluble E-selectin in patients with IDDM and NIDDM: relation to metabolic control. *Diabetologia* 1995; 38: 1122-1124.

[11] Davies MJ, Gordon JL, Gearing AJH, Pigott R, Woolf N, Katz D, Kyriakopoulos A. The expression of the adhesion molecules ICAM-1, VCAM-1, PECAM-1, and E-selectin in human atherosclerosis. *J Pathol* 1993; 171: 223-229.

[12] Cattell V. Macrophages in acute glomerular inflammation. *Kidney Int* 1994; 45:945-952.

[13] Ihm CG, Park JK, Hong SP, Lee TW, Cho BS, Kim MJ, Cha DR, Ha H. A high glucose concentration stimulates the expression of monocyte chemotactic peptide 1 in human mesangial cells. *Nephron* 1998; 79: 33-37.

[14] Nomura S, Shouzu A, Omoto S, Nishikawa M, Fukuhara S. Significance of chemokines and activated platelets in patients with diabetes. *Clin Exp Immunol* 2000; 121: 437-443.

[15] Koch AE, Kunkel SL, Harlow LA, Mazarakis DD, Haines GK, Burdick MD, Pope RM, Stieter RM. Macrophage inflammatory protein-1: a novel chemokine cytokine for macrophages in rheumatoid arthritis. *J Clin Invest* 1994; 93: 921-928.

[16] Pattison J, Nelson PJ, Huie P, von Leuttichau I, Farshid G, Sibley RK, Krensky AM. RANTES chemokine expression in cell-mediated transplant rejection of the lidney. *Lancet* 1994; 343: 209-211.

[17] Cocchi F, DeVico AL, Garzino-Demo A, Arya SK, Gallo RC, Lusso P. Identification of RANTES, MIP-1α , and MIP-1β as the major HIV-suppressive factors produced by $CD8^+$ T cells. *Science* 1995; 270: 1811-1815.

[18] Holme PA, Muller F, Solum NO, Brosstad F, Froland SS, Aukrust P. Enhanced activation of platelets with abnormal release of RANTES in HIV-1 infection. *FASEB J* 1998; 12: 79-90.

[19] Weyrich AS, Elstad MR, McEver RP, McIntype TM, Moore KL, Morrisey JH, Prescott SM, Zimmerman GA. Activated platelets signal chemokine synthesis by human monocytes. *J Clin Invest* 1996; 97: 1525-1534.

[20] Nomura S, Shouzu A, Omoto S, Nishikawa M, Iwasaka T. Effects of losartan and simvastatin on monocyte-derived microparticles in hypertensive patients with and without type 2 diabetes mellitus. *Clin Appl Thromb Hemost* 2004; 10: 133-141.

[21] Nomura S, Shouzu A, Omoto S, Nishikawa M, Iwasaka T. Benidipine imprives oxidized LDL-dependent monocyte and endothelial dysfunction in hypertensive patients with type 2 diabetes mellitus. *J Hum Hypertens* 2005; 19: 551-557.

[22] Nomura S, Shouzu A, Omoto S, Nishikawa M, Iwasaka T. Long-term treatment with nifedipine modulates procoagulant marker and C-C chemokine in hypertensive patients with type 2 diabetes mellitus. *Thromb Res* 2005; 115: 277-285.

[23] Daae LN, Kierulf P, Landass S, Urdal P. Cardiovascular risk factors: interactive effects of lipids, coagulation and fibrinolysis. *Scand J Clin Lab Invest* 1993; 215: 19-27.

[24] Fitzgerald DJ, Roy L, Catella F, Fitzgerald GA. Platelet activation in unstable coronary disease. *N Engl J Med* 1986; 315: 983-989.

[25] Ernst E, Hammerschmidt DE, Bagge U, Matrai A, Dormandy JA. Leukocytes and the risk of ischemic diseases. *JAMA* 1987; 257: 2318-2324.

[26] Leatham EW, Bath PM, Tooze JA, Camm AJ. Increased monocyte tissue factor expression in coronary disease. *Br Heart J* 1995; 73: 10-13.

[27] Schwager I, Jungi TW. Effect of recombinant cytokines on the induction of macrophage procoagulant activity. *Blood* 1994; 83: 152-161.

[28] Gawaz M, Neumann FJ, Ott I, Schiessler A, Schomig A. Platelet function in acute myocardial infarction treated with direct angioplasty. *Circulation* 1996; 93: 229-237.

[29] Nomura S. Function and clinical significance of platelet-derived microparticles. *Int J Hematol* 2001; 74: 397-404.

[30] Matsumoto N, Nomura S, Kamihata H, Kimura Y, Iwasaka T. Association of platelet-derived microparticles with C-C chemokines on vascular complication in patients with acute myocardial infarction. *Clin Appl Throm Hemost* 2002; 8: 279-286.

[31] Nomura S, Uehata S, Saito S, Osumi K, Ozeki Y, Kimura Y. Enzyme immunoassay detection of platelet-derived microparticles and RANTES in acute coronary syndrome. *Thromb Haemost* 2003; 89: 506-512.

[32] Cines DB, Blanchette VS. Immune thrombocytopenic purpura. *N Engl J Med* 2002; 346: 995-1008.

[33] Emilia G, Longo G, Luppi M, Gandini G, Morselli M, Ferrara L, Amarri S, Cagossi K, Torelli G. Helicobacter pylori eradication can induce platelet recovery in idiopathic thrombocytopenic purpura. *Blood* 2001; 97: 812-814.

[34] Banford KB, Andersen L. Host response. *Current Opinions in Gastroenterology* 1997; 13: 25-30.

[35] Nagahama M, Nomura S, Kanazawa S, Ozaki Y, Kagawa H, Fukuhara S. Significance of chemokines and soluble CD40 ligand in patients with autoimmune thrombocytopenic purpura. *Eur J Haematol* 2002; 69: 303-308.

[36] Watanabe N, Shimada T, Ohtsuka Y, Hiraishi H, Terano A. Proinflammatory cytokines and Helicobacter pylori stimulate CC-chemokine expression in gastric epithelial cells. *J Physiol Pharmacol* 1997; 48: 405-413.

[37] Innocenti M, Svennerhorm AM, Quinding-Jarbrink M. Helicobacter pylori lipopolysaccharides preferentially induce CXC chemokine production in human monocytes. *Infect Immunol* 2001; 69: 3800-3808.

[38] Sugiura T, Kawaguchi Y, Harigai M, Takagi K, Ohta S, Fukasawa C, Hara M, Kamatani N. Increased CD40 expression on muscle cells of polymyositis and dermatomyositis: role of CD40-CD40 ligand interaction in IL-6, IL-8, IL-15, and monocyte chemoattractant protein-1 production. *J Immunol* 2000; 164: 6593-6600.

[39] Jarque I, Andreu R, Lopis I, De la Rubia J, Gomis F, Senent L, Jimenez C, Martin G, Martinez JA, Sanz GF, Ponce J, Sanz MA. Absence of platelet response after eradication of Helicobacter pylori infection in patients with chronic idiopathic thrombocytopenic purpura. *Br J Haematol* 2001; 115: 1002-1003.

[40] Michel M, Cooper N, Jean C, Frissora C, Bussel JB. Does Helicobacter pylori initiate or perpetuate immune thrombocytopenic purpura ? *Blood* 2004; 103: 890-896.

[41] Roberts MM, To LB, Gillis D, Mundy J, Rawing C, Ng K, Juttner CA. Immune reconstitution following peripheral blood stem cell transplantation, autologous bone marrow transplantation and allogenic bone marrow transplantation. *Bone Marrow Transplant* 1993; 12: 469-475.

[42] Weinthal JA. The role of cytokines following bone marrow transplantation: indications and controversies. *Bone Marrow Transplant* 1996; 18(Suppl. 3): S10-14.

[43] Ogawa M. Differentiation and proliferation of hematopoietic stem cells. *Blood* 1993; 81: 2844-2853.

[44] Holler E, Kolb HJ, Moller A, Kempeni J, Liesenfeld S, Pechumer H, Lehmacher W, Ruckdeschel G, Gleixner B, Riedner C, Ledderose G, Brehm G, Mittermuller J, Wilmanns W. Increased serum levels of tumor necrosis factor α precede major complications of bone marrow transplantation. *Blood* 1990; 75: 1011-1016.

[45] Rowbottom AW, Riches PG, Downie C, Hobbs JR. Monitoring cytokine production in peripheral blood during acute graft-versus-host disease following allogenic bone marrow transplantation. *Bone Marrow Transplant* 1993; 12: 635-641.

[46] Imamura M, Hashino S, Kobayashi H, Kubayashi S, Hirano S, Minagawa T, Tanaka J, Fujii Y, Kobayashi M, Kasai M. Serum cytokine levels in bone marrow transplantation: synergistic interaction of interleukin-6, interferon-gamma, and tumor necrosis factor-alpha in graft-versus- host disease. *Bone Marrow Transplant* 1994; 13: 745-751.

[47] Testa S, Manna A, Porcellini A, Maffi F, Morstabilini G, Denti N, Macchi S, Rosti G, Porcellini G, Cassi D, Ferrari L. Increased plasma level of vascular endothelial glycoprotein thrombomodulin as an early indicator of endothelial damage in bone marrow transplantation. *Bone Marrow Transplant* 1996; 18: 383-388.

[48] Nomura S, Ishii K, Kanazawa S, Inami N, Kamitsuji Y, Uoshima N, Ishida H, Yoshihara T, Kitayama H, Hayashi K. Role of platelet-derived chemokines (RANTES and ENA-78) after stem cell transplantation. *Transplant Immunol* 2006; 15: 247-253.

[49] Nasu K, Arima K, Kai K, Fujisawa K, Nishida M, Miyakawa I. Expression of epithelial neutrophil-activating peptide 78 in cultured human endometrial stromal cells. *Molecular Human Reproduction* 2001; 7: 453-458.

[50] Power CA, Furness RB, Brawand C, Wells TN. Cloning of a full-length cDNA encoding the neutrophil-activating peptide ENA-78 from human platelets. *Gene* 1994; 151: 333-334.

[51] Clemetson KJ, Clemetson JM, Proudfoot AEI, Power CA, Baggiolini M, Wells TN. Functional expression of CCR1, CCR3, CCR4, and CXCR4 chemokine receptors on human platelets. *Blood* 2000; 96: 4046-4054.

[52] Schots R, Kaufman L, Van Riet I, Ben Othman T, De Waele M, Van Camp B, Demanet C. Proinflammatory cytokines and their role in the development of major transplant-related complications in the early phase after allogenic bone marrow transplantation. *Leukemia* 2003; 17: 1150-1156.

[53] Grimm J, Zeller W, Zander AR. Soluble interleukin-2 receptor serum levels after allogenic bone marrow transplantations as a marker for GVHD. *Bone Marrow Transplant* 1998; 21: 29-32.

[54] Rus V, Svetic A, Nguyen P, Gause WC, Via CS. Kinetics of TH1 and Th2 cytokine production during the early course of acute and chronic murine graft-versus-host disease. *J Immunol* 1995; 155: 2396-2406.

[55] Murai M, Yoneyama H, Harada A, Yi Z, Verstegaard C, Guo B, Suzuki K, Asakura H, Matsushima K. Active participation of $CCR5^{+}CD8^{+}$ T lymphocytes in the pathogenesis of liver injury in graft-versus-host disease. *J Clin Invest* 1999; 104: 49-57.

[56] Zhang Y, McCormick LL, Desai SR, Wu C, Gillian AC. Murine sclerodermatous graft-versus-host disease, a model for human scleroderma: cutaneous cytokines, chemokines, and immune cell activation. *J Immunol* 2002; 168: 3088-3098.

[57] Kawai T, Seki M, Hiromatsu K, Eastcott JW, Watts GF, Sugai M, Smith DJ, Porcelli SA, Taubman MA. Selective diapedesis of Th1 cells induced by endothelial cell RANTES. *J Immunol* 1999; 163: 3269-3278.

[58] Min CK, Lee MY, Min DJ, Lee DG, Kim YJ, Park YH, Kim HJ, Lee S, Kim DW, Lee JW, Min WS, Kim CC. The kinetics of circulating cytokines including IL-6, TNF-α, IL-8 and IL-10 following allogenic hematopoietic ctem cell transplantation. *Bone Marrow Transplant* 2001; 28: 935-940.

[59] Lee WY, Baek KH, Rhee EJ, Tae HJ, Oh KW, Kang MI, et al. Impact of circulating bone-resorbing cytokines on the subsequent bone loss following bone marrow transplantation. *Bone Marrow Transplant* 2004; 34: 89-94.

[60] Nomura S, Inami N, Kanazawa S, Iwasaka T, Fukuhara S. Elevation of platelet activation markers and chemokines during peripheral blood stem cell harvest with G-CSF. *Stem Cells* 2004; 22: 696-703.

[61] Kanazawa S, Muramatsu M, Kinoshita Y, Yamaguchi K, Nomura S. Gefitinib has the potential of activating cell immunity against malignant cells. *J Clin Oncol* 2005; 23: 3865-3866.

[62] Kanazawa S, Yamaguchi K, Kinoshita Y, Komiyama Y, Muramatsu M, Nomura S. Elevation of soluble interleukin-2 receptor in patients with non-small cell lung cancer treated with gefitinib. *J Cancer Res Clin Oncol* 2006; 132: 719-725.

In: Chemokine Research Trends
Editor: Leah R. Grinwald, pp. 171-180

ISBN: 978-60021-588-9

Chapter 12

The Role of the CXCL16-CXCR6 Pathway in Cell Recruitment, Activation and Effector Functions

Gunilla A. Lundberg
Örebro University, Dept of Clinical Medicine, Biomedicine, S-70182 Örebro, Sweden

Abstract

The recently discovered chemokine/scavenger receptor CXCL16 is an interesting chemokine molecule, which is emerging as a highly significant player in atherosclerosis, rheumatoid arthritis (RA), neuroinflammatory diseases and other conditions.

CXCL16 was first identified as a scavenger receptor for oxidized LDL (oxLDL). It was shown to play a role in the internalization of oxLDL particles, leading to subsequent foam cell formation, which is a critical step in the development of atherosclerotic plaques. It is a transmembrane molecule, which is expressed in dendritic cells, macrophages and aortic smooth muscle cells and isaos cleaved by the metalloprotease ADAM-10 to release a soluble form that has chemotactic activity and is the only known ligand of the chemokine receptor CXCR6 which is expressed on a subset of T cells and NKT cells. CXCL16 may therefore play an important role in enhancing effector functions of T-cells.

CXCL16 has been detected in human and murine atherosclerotic lesions, and elevated levels of CXCL16 in cerebrospinal fluid suggest local production of CXCL16 in the central nervous system. It is also expressed in the synovium of RA patients. This chapter identifies important roles for CXCL16 in T-cell triggering and activation of effector functions, and describes research on understanding its significance in atherosclerotic vascular disease, immune responses in the central nervous system and the infiltration of leukocytes into synovial tissue and fluid in patients with the chronic inflammatory disease RA.

Introduction

Chemokines and their receptors have long been recognized as key molecules directing leukocyte migration between blood, lymph and tissues. Evidence has accumulated in recent years indicating that, in addition to their chemotactic functions, chemokine receptors are highly versatile players that fine-tune immune responses. Chemokine receptors and ligands have been implicated in dendritic-cell (DC) maturation, signal transmission at the immunological synapse between T lymphocytes and their cellular partners, and in the polarization of immune responses. These findings identify new roles for chemokines in T-cell triggering and activation of effector functions, and suggest that these small cytokines represent key conductors of adaptive immunity [Charo et al 2006, Sheikne & Hansson 2004]. Chemokines and their receptors have been implicated in a wide range of diseases with prominent inflammatory components. In this chapter studies on the recently discovered chemokine CXCL16 and its receptor CXCR6 will be described and the importance of their role in selected clinical disorders discussed.

Chemokine CXCL16

The chemokine CXCL16 was independently cloned as a scavenger receptor for phosphatidylserine and oxidized lipid (SR-PSOX) and as a chemokine (CXCL16) [Shimaoka et al 2000, Matloubin et al 2000, Wilbanks et al 2001]. It is expressed on macrophages and subtypes of DCs and, at lower levels, on T cells [Tabata et al 2005]. This chemokine is now emerging as an important player in atherosclerosis, rheumatoid arthritis, neuroinflammatory diseases and other conditions

In normal circumstances, active soluble chemokines are rapidly sequestered by sulphated glycosaminoglycan (GAG) components of proteoglycans on the cell-surface or within the extracellular matrix. This binding typically occurs through interaction between characteristic basic amino acid domains and heavily sulphated, anionic sequences within heparin sulphate [Ali et al 2002, Carter et al 2003]. The interaction of chemokines and GAGs on endothelial surfaces is a crucial step for establishing a chemotactic gradient and for subsequent presentation of active chemokines to their G-protein coupled receptors (GPCR). In order to provide the vectorial cues necessary to drive leukocyte migration towards the site of chemokine production, it is necessary for chemokines to produce stable concentration gradients; these are maintained by interactions between chemokines and their supporting proteoglycans [Proudfoot et al 2003, Ali et al 2005].

Although most chemokines are present in soluble forms and some may be associated with GAG moieties on the cell surface, two of the chemokines, namely CX3CL1 (fractalkine) and CXCL16, have a natural mucin stalk that adheres onto the membrane of the cells that produce them [Bazan et al 1997, Matloubin et al 2001]. Their "chemokine" domain is located at the N-terminus of the mucin stalk and can be released by metalloprotease cleavage. While the soluble, released chemokine domain of CX3CL1 and CXCL16 functions similarly to other secreted chemokines, their membrane bound forms play an important role in mediating leukocyte-endothelial cell adhesion and extravasation.

CXCL16 Signalling Through the Chemokine Receptor CXCR6

The chemokines produce their effects through activation of the seven-transmembrane domain of GPCR, so called chemokine receptors. While several chemokine receptors are promiscuous, binding multiple ligands with similar high affinity, some bind only a single ligand. The CXCR6 receptor is of the single ligand-binding type and binds only CXCL16. CXCL16 can capture cells bearing the CXCR6 receptor [Nakayama et al 2003, Shimaoka et al 2004], and in the process is cleaved by the metalloprotease ADAM-10 to release a soluble form that has chemotactic activity [Abel et al 2004, Gough et al 2004]. Activation of chemokine receptors by their ligand, typically leads to rapid internalisation and inactivation [Signoret et al 1997].

CXCR6 is expressed on a subset of T cells and natural killer T (NKT) cells but functional CXCR6 is apparently absent from naïve T cell in humans. Regulation of CXCR6 on lymphocytes parallels that of CCR5 and is inversely correlated with CXCR4 [Johnston et al 2003]. Kim *et al* demonstrated that CXCR6 expression is upregulated in T cells by interaction with DCs and that addition of IL-12 enhances the generation of CXCR6 positive cells whereas IL-4 reduces it. Furthermore, most of the CXCR6 positive cells in blood are IFN-γ secreting T helper T-cells, cytotoxic T cells, or CD56 positive T cells that are enriched for effector functions and IFN-γ release. Thus upregulation of CXCR6 on primed T cells by DCs in an IL-12 rich environment will lead to the release of effector cells with the potential to mediate type I responses. In order to complete the story, the authors describe enrichment of CXCR6 cells in inflamed tissue, such as rheumatoid joints and inflamed livers, implying the receptor is involved in effector cell recruitment to (or retention in) tissue [Kim et al 2001, Johnston et al 2003].

Direct contacts between DCs and T cells or NKT cells play important roles in primary and secondary responses. CXCL16 functions as a direct adhesion cell molecule for cells expressing its receptor CXCR6 [Shimaoka et al 2004]. After binding to CXCR6, CXCL16 activates the expression of the nuclear factor kappa B (NF-κB)-mediated tumor necrosis factor (TNF)-α [Matloubin et al 2000]. CXCL16-mediated NF-κB activation occurs via heterotrimeric G proteins, phosphoinositide 3-kinase (PI3K), 3-phosphoinositide-dependent kinase-1 (PDK-1), protein kinase B (PBK/Akt), and IκB kinase (IKK) and as a part of this process CXCL16 can induce IκB phosphorylation and degradation [Chandrasekar et al 2004].

CXCL16-CXCR6 IN ATHEROSCLEROSIS

CXCL16 has been implicated in the development of atherosclerosis. Its expression is prominent in human atherosclerotic lesions, where it can be up-regulated by the pro-atherogenic cytokine interferon-γ (Minami et al 2001b, Wuttge et al 2004). Expression of endogenous CXCL16 is induced by IFN-γ and TNF-α in endothelial cells (ECs) and vascular smooth muscle cells [Abel et al. 2004, Wågsäter et al. 2004]. Using ADAM10-deficient fibroblasts, Abel *et al* (2004) showed that endogenous ADAM10 is the most relevant sheddase of endogenous CXCL16.

After binding to the specific chemokine receptor CXCR6, CXCL16 activates nuclear factor kappa B (NF-κB)-mediated TNF-α expression and induces cell proliferation in smooth muscle cells [Chandrasekar et al 2004], processes that contribute to the development of atherosclerotic/restenotic lesions.

In human atherosclerotic lesions, Minami *et al.* (2001) have localized CXCL16 expression predominantly to lipid-laden macrophages accumulated in the intima. In contrast, normal arterial wall failed to show CXCL16 immunoreactivity. However, in cultured aortic smooth muscle cells (ASMC), Hofnagel *et al.* (2002) have demonstrated CXCL16 expression by RT-PCR. Together, these reports indicate that CXCL16 is expressed in atherosclerotic lesions and that both macrophages and ASMC express CXCL16, and suggest that CXCL16 may play a role in the pathobiology of atherosclerosis. Human ASMC express CXCR6 and treatment with CXCL16 activates NF-κB. Most importantly, treatment of ASMC with CXCL16 increases cell-cell adhesion and cellular proliferation in a NF-κB-dependent manner [Chandrasekar et al 2004].

In these experiments by Chandrasekar *et al* (2004), human ASMC were treated with pharmacological inhibitors or transiently transfected with pathway-specific dominant-negative or kinase-dead expression vectors prior to the addition of CXCL16. Human ASMC expressed CXCR6 at basal conditions and exposure of human ASMC to CXCL16 increased NF-κB DNA binding activity, induced κB-driven luciferase activity, and up-regulated TNF-α expression in an NF-κB-dependent manner. However, treatment with pertussis toxin (a G_i inhibitor), a PI3K, or a PBK/Akt inhibitor significantly attenuated CXCL16-induced NF-κB activation. The same results were seen if there was overexpression of dominant-negative PI3Kγ, dnPDK-1, kinase-dead Akt, kdIKK-β, dnIKK-γ, dnIκB-α, or dnIκB-β [Chandrasekar et al 2004].

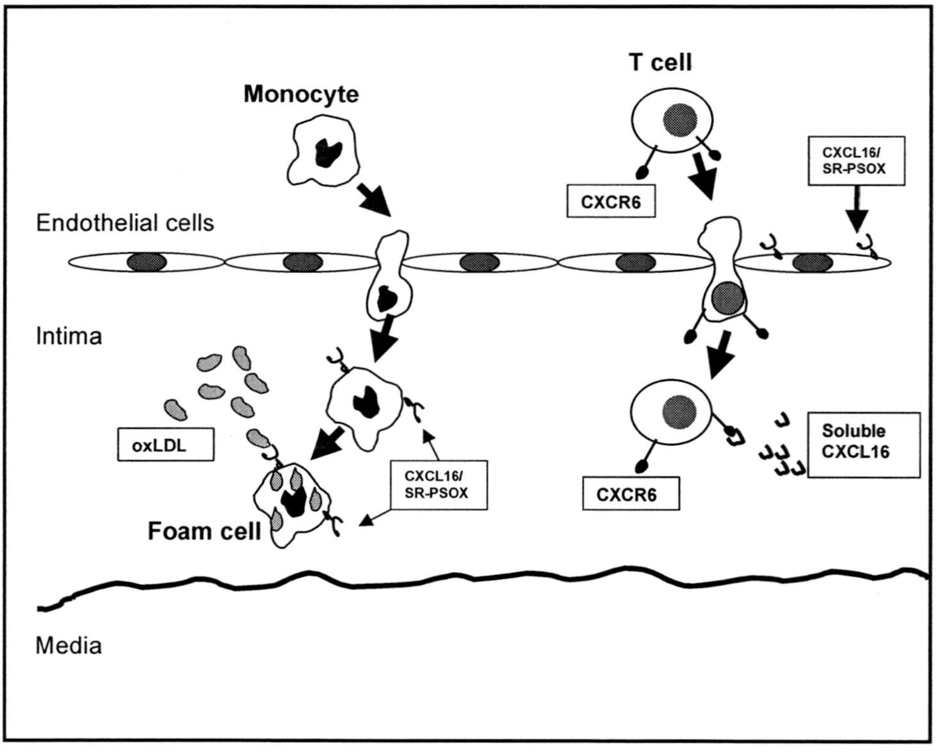

Figure 1. The role of CXCL16 and CXCR6 in vascular inflammation and atherosclerosis. The schematic shows a monocyte migrating into the arterial intima where it aquires the morphological characteristics of a macrophage. This expresses membrane-bound CXCL16 molecules which function as scavanger receptors for oxLDL and in the process give rise to a macrophage foam cell. The figure also illustrates recruitment of a CXCR6-positive T cell into the intima by chemotaxis coupled with direct cell-cell adhesion to membrane-bound and soluble CXCL16 .

CXCL16 is a potent and direct activator of NF-κB and induces κB-dependent proinflammatory gene transcription. Most importantly, CXCL16 increases cell-cell adhesion and induces κB-dependent ASMC proliferation, indicating that CXCL16 may play an important role in the development and progression of atherosclerotic vascular disease. Figure 1 shows the role of CXCL16 and CXCR6 in vascular inflammation and atherosclerosis.

THE RELATIONSHIP BETWEEN CXCL16-CXCR6 AND CORONARY ARTERY DISEASE

The relationship between plasma CXCL6 levels and coronary artery diseases (CAD) has been studied in a group of healthy individuals and patients with different types of CAD [Sheikine et al 2006]. Patients with different CAD were found to exhibit lower systemic CXCL16 levels compared with controls, while CXCL16 levels were not related to the degree of coronary artery stenosis or to biochemical risk indicators of CAD. CXL16 cleavage from the surface of ECs and SMCs [Abel et al 2004] following the inflammatory activation that is observed in CAD can potentially lead to changes in the circulating CXCL16 concentrations among CAD patients. Based on this data the authors hypothesize that soluble CXCL16 acts as a scavenger for oxidized LDL (oxLDL).

A recent study in the laboratory of the author showed an association between a CXCL16 polymorphism and the severity of coronary artery stenosis. This work investigated the possibility of a correlation of CXCL16 expression and coronary artery disease in two human patient cohorts. In the first cohort, the effect of an alanine-to-valine polymorphism of CXCL16 (which is postulated to influence the release of CXCL16 from the cell surface) was examined in post-myocardial infarction subjects [Lundberg at al 2005]. The results showed that carriers of the polymorphism had a higher severity of coronary artery stenosis.

These patient studies further underscore the link between CXCL16 and coronary artery disease and indicate that CXCL16 may be atheroprotective in the context of human disease. This might also be of interest in chronic inflammatory conditions, in which a role for CXCL16 in the pathogenesis has been suggested. Interestingly, patients with chronic inflammatory conditions have a significantly increased incidence of cardiovascular disease and related morbidity and mortality. In fact, patients with rheumatoid arthritis (RA) have an increased risk of developing CAD, which is independent of disease activity [Boers M 2004]. Intriguingly, CXCL16 was recently found to be increased in synovial fluid and tissue of RA patients [van der Voort et al, 2005, Ruth et al 2006].

CXCL16-CXCR6 IN RHEUMATOID ARTHRITIS

Rheumatoid arthritis is a chronic inflammatory disease that affects synovial tissue in multiple joints. The inflammatory process is characterized by infiltration of leukocytes into the synovial tissue and the synovial fluid in the joints. This is achieved by the leukocytes adhering to the activated endothelium, leading to migration across the endothelial barrier into the synovial tissue [Springer 1994]. Investigators have recently extended initial observational studies to explore the role of chemokines and chemokine receptors in the pathogenisis of inflammatory joints.

Kim *et al* (2001) described expression of CXCR6 in a small number of RA patients and reported an accumulation of CXCR6 positive T cells in their synovial fluid. In a recent study by van der Voort *et al* (2005) of joints from RA patients and healthy individuals there was found to be strongly enhanced expression of CXCL16 and its characterized protease ADAM-10 in RA synovia, which resulted in the recruitment of CXCR6 positive memory cells in RA joints. Nanki *et al* (2005) examined the role of CXCL16 and CXCR6 in T cell migration and stimulation in RA synovium and showed that CXCL16 was active in the synovial fluid of people with RA. They also showed participation of CXCL16 and its cellular receptor CXCR6 in leukocyte migration to the synovial tissue in RA and found that CXCL16 was expressed in RA synovium. Not only was CXCR6 expressed more frequently on synovial T cells than in peripheral blood but CXCR6-positive synovial T cells also expressed activated markers of inflammation CD154 and the inducible costimulator (ICOS) more than did CXCR6-negative T cells. They were also able to halt the signalling cascade and the migratory pattern of leukocytes to the inflamed tissue by blocking the intracellular pathways used by CXCL16. Taken together, these results indicate that CXCL16 plays an important role in T cell accumulation and stimulation in synovium and suggest that CXCL16 could be a target molecule in new therapies for RA.

Consistant with these conclusions, Ruth *et al* (2006) have recently shown intense macrophage and lining cell staining for CXCL16 in RA synovial tissue which correlated with increased CXCL16 messenger levels compared to normal synovial tissue. Moreover, Kageyana *et al* (2006) have very recently shown that treatment with the anti-TNFα antibody infliximab lowered the serum levels of CXCL16 in 23 patients with RA studied over a period of 30 weeks, indicating that CXCL16 may be one of the chemokines regulated by infliximab treatment.

CXCL16-CXCR6 in Inflammatory Disorders of CNS

The endogenous chemokine system in the brain may act in concert with the neurotransmitter and neuropeptide systems to govern brain function. It can thus be thought of as the third major system in the brain [Adler et al 2006].

Various studies have shown that microglia are able to perform as antigen presenting cells and proinflammatory effector functions following activation, and are capable of producing a wide array of chemokines [Aloisi 2001]. Participation of the chemokine system in normal CNS homeostasis and in pathological states such as multiple sclerosis (MS) is not fully elucidated. MS is an inflammatory, demyelinating disorder of CNS, that commonly results in axonal degeneration and progressive disability. A potential role for CXCL16 was recently suggested in experimental auto-immune encephalomyelitis, an animal model of MS [Fukomoto et al 2004].

Levels of CXCL16 have been found to be significantly elevated in cerebrospinal fluid compared to serum and, interestingly, CXCL16 levels both in serum and cerebrospinal fluid (CSF) are significantly elevated in different inflammatory conditions as systemic lupus erythematosus and MS [le Blanc et al 2006]. These findings suggest an important role for CXCL16 in the attraction of activated T cells into the brain. The elevated levels of CXCL16 in cerebrospinal fluid suggest local production of CXCL16 in the CNS. Another explanation for the high CXCL16 levels might be an enhanced cleavage by proteases. Evidence for this is that the protease ADAM-10, which is able to cleavage membrane-bound CXCL16, was recently found to be expressed constitutively in both normal and inflamed CNS [Kieseier et al. 2003].

CXCL16 is highly expressed in human gliomas, while expression in normal brain is low and mainly restricted to brain vascular endothelial cells [Ludwig et al 2005]. The authors demonstrated that CXCL16 is expressed by endothelial cells, astrocytes and glioma cells *in situ* and that besides T cells and microglial cells, activated astrocytes were also found to express CXCR6 mRNA. Astroglial and microglial cells may therefore constitute a target for autocrine or paracrine stimulation by CXCL16. Glial cells themselves were attracted by CXCL16, which might favour interaction between glial cells and activated T cells, leading to an even more pronounced immune response. Given the fact that macrophages are the main producers of CXCL16, microglia and/or astrocytes are the most likely candidates responsible for the production of CXCL16 in the CSF as microglia are the resident mononuclear phagocytes within the CNS. Consistent with this hypothesis is the finding that glial CXCL16 expression is further enhanced by INF-γ and TNF-α [Ludwig et al 2005].

Conclusion

CXCL16 is a recently identified chemokine and scavenger receptor with characteristics that implicate it in the process of inflammation. It is, for example, proatherogenic and is expressed in atherosclerotic lesions. Its expression is strongly enhanced in joints of patients with RA compared to healthy individuals and levels of CXCL16 are significantly elevated in CSF in different inflammatory conditions of the brain. Moreover, CXCL16 is shown to attract CXCR6 positive cells to the site of inflammation.

CXCL16 and its receptor CXCR6, by virtue of their potential role in the mediation of inflammation in atherosclerosis, RA and MS, provide a potentially very important site for pharmacological intervention. The recent work of Kageyana *et al* (2006) demonstrating that treatment with the anti-TNFα antibody infliximab lowered the serum levels of CXCL16 in 23 patients with RA, is an indication of one possible approach toward attempting such interventions. Chemokine ligand-receptor interactions, however, can be modulated in several ways, among these being the reduction of signal-transduction mechanisms that are required for chemokine synthesis and disruption of chemokine ligand and receptor interaction.

It is hoped that further understanding of the molecular biology of CXCL16 and its receptor CXCR6 in health and disease will reveal the potential for further valuable roles in therapeutic intervention.

References

Abel S, Hundhausen C, Mentlein R, Schulte A, Berkhout TA, Broadway N, Hartmann D, Sedlacek R, Dietrich S, Muetze B, Schuster B, Kallen KJ, Saftig P, Rose-John S, Ludwig A. The transmembrane CXC-chemokine ligand 16 is induced by IFN-gamma and TNF-alpha and shed by the activity of the disintegrin-like metalloproteinase ADAM10. *J Immunol.* 2004 May 15;172(10):6362-72.

Ali S, Fritchley SJ, Chaffey BT, Kirby JA. Contribution of the putative heparan sulfate-binding motif BBXB of RANTES to transendothelial migration. *Glycobiology.* 2002 Sep;12(9):535-43.

Ali S, Robertson H, Wain JH, Isaacs JD, Malik G, Kirby JA. A non-glycosaminoglycan-binding variant of CC chemokine ligand 7 (monocyte chemoattractant protein-3) antagonizes chemokine-mediated inflammation. J Immunol. 2005 Jul 15;175(2):1257-66.

Aloisi F. Immune function of microglia. *Glia.* 2001 Nov;36(2):165-79.

Aslanian AM, Charo IF. Targeted disruption of the scavenger receptor and chemokine CXCL16 accelerates atherosclerosis. *Circulation.* 2006 Aug 8;114(6):583-90.

Bazan JF, Bacon KB, Hardiman G, Wang W, Soo K, Rossi D, Greaves DR, Zlotnik A, Schall TJ. A new class of membrane-bound chemokine with a CX3C motif. *Nature.* 1997 Feb 13;385(6617):640-4.

Boers M, Dijkmans B, Gabriel S, Maradit-Kremers H, O'Dell J, Pincus T. Making an impact on mortality in rheumatoid arthritis: targeting cardiovascular comorbidity. *Arthritis Rheum.* 2004 Jun;50(6):1734-9.

Carter NM, Ali S, Kirby JA. Endothelial inflammation: the role of differential expression of N-deacetylase/N-sulphotransferase enzymes in alteration of the immunological properties of heparan sulphate. *J Cell Sci.* 2003 Sep 1;116(Pt 17):3591-600.

Chandrasekar B, Bysani S, Mummidi S. CXCL16 signals via Gi, phosphatidylinositol 3-kinase, Akt, I kappa B kinase, and nuclear factor-kappa B and induces cell-cell adhesion and aortic smoothmuscle cell proliferation. *J Biol Chem.* 2004 Jan 30;279(5):3188-96.

Charo IF, Ransohoff RM. The many roles of chemokines and chemokine receptors in inflammation. *N Engl J Med.* 2006 Feb 9;354(6):610-21.

Fukumoto N, Shimaoka T, Fujimura H, Sakoda S, Tanaka M, Kita T, Yonehara S. Critical roles of CXC chemokine ligand 16/scavenger receptor that binds phosphatidylserine and oxidized lipoprotein in the pathogenesis of both acute and adoptive transfer experimental autoimmune encephalomyelitis. *J Immunol.* 2004 Aug 1;173(3):1620-7.

Gough PJ, Garton KJ, Wille PT, Rychlewski M, Dempsey PJ, Raines EW. A disintegrin and metalloproteinase 10-mediated cleavage and shedding regulates the cell surface expression of CXC chemokine ligand 16. *J Immunol.* 2004 Mar 15;172(6):3678-85.

Hofnagel O, Luechtenborg B, Plenz G, Robenek H. Expression of the novel scavenger receptor SR-PSOX in cultured aortic smooth muscle cells and umbilical endothelial cells.*Arterioscler Thromb Vasc Biol.* 2002 Apr 1;22(4):710-1.

Johnston B, Kim CH, Soler D, Emoto M, Butcher EC. Differential chemokine responses and homing patterns of murine TCR alpha beta NKT cell subsets. *J Immunol.* 2003 Sep 15;171(6):2960-9.

Kieseier BC, Pischel H, Neuen-Jacob E, Tourtellotte WW, Hartung HP. ADAM-10 and ADAM-17 in the inflamed human CNS. *Glia.* 2003 Jun;42(4):398-405.

Kim CH, Kunkel EJ, Boisvert J, Johnston B, Campbell JJ, Genovese MC, Greenberg HB, Butcher EC. Bonzo/CXCR6 expression defines type 1-polarized T-cell subsets with extralymphoid tissue homing potential. *J Clin Invest.* 2001 Mar;107(5):595-601.

le Blanc LM, van Lieshout AW, Adema GJ, van Riel PL, Verbeek MM, Radstake TR. CXCL16 is elevated in the cerebrospinal fluid versus serum and in inflammatory conditions with suspected and proved central nervous system involvement. *Neurosci Lett.* 2006 Apr 10-17;397(1-2):145-8.

Ludwig A, Hundhausen C, Lambert MH, Broadway N, Andrews RC, Bickett DM, Leesnitzer MA, Becherer JD. Metalloproteinase inhibitors for the disintegrin-like metalloproteinases ADAM10 and ADAM17 that differentially block constitutive and phorbol ester-inducible shedding of cell surface molecules. *Comb Chem High Throughput Screen.* 2005 Mar;8(2):161-71.

Lundberg GA, Kellin A, Samnegard A, Lundman P, Tornvall P, Dimmeler S, Zeiher AM, Hamsten A, Hansson GK, Eriksson P. Severity of coronary artery stenosis is associated with a polymorphism in the CXCL16/SR-PSOX gene. *J Intern Med.* 2005 May;257(5):415-22.

Matloubian M, David A, Engel S, Ryan JE, Cyster JG. A transmembrane CXC chemokine is a ligand for HIV-coreceptor Bonzo. *Nat Immunol.* 2000 Oct;1(4):298-304.

Minami M, Kume N, Shimaoka T, Kataoka H, Hayashida K, Yonehara S, Kita T. Expression of scavenger receptor for phosphatidylserine and oxidized lipoprotein (SR-PSOX) in human atheroma. *Ann N Y Acad Sci.* 2001 Dec;947:373-6.

Minami M, Kume N, Shimaoka T, Kataoka H, Hayashida K, Akiyama Y, Nagata I, Ando K, Nobuyoshi M, Hanyuu M, Komeda M, Yonehara S, Kita T. Expression of SR-PSOX, a novel cell-surface scavenger receptor for phosphatidylserine and oxidized LDL in human atherosclerotic lesions. *Arterioscler Thromb Vasc Biol.* 2001 Nov;21(11):1796-800.

Nakayama T, Hieshima K, Izawa D, Tatsumi Y, Kanamaru A, Yoshie O. Cutting edge: profile of chemokine receptor expression on human plasma cells accounts for their efficient recruitment to target tissues. *J Immunol.* 2003 Feb 1;170(3):1136-40.

Nanki T, Shimaoka T, Hayashida K, Taniguchi K, Yonehara S, Miyasaka N. Pathogenic role of the CXCL16-CXCR6 pathway in rheumatoid arthritis. *Arthritis Rheum.* 2005 Oct;52(10):3004-14.

Proudfoot AE, Handel TM, Johnson Z, Lau EK, LiWang P, Clark-Lewis I, Borlat F, Wells TN, Kosco-Vilbois MH. Glycosaminoglycan binding and oligomerization are essential for the in vivo activity of certain chemokines. *Proc Natl Acad Sci U S A.* 2003 Feb 18;100(4):1885-90.

Ruth JH, Haas CS, Park CC, Amin MA, Martinez RJ, Haines GK 3rd, Shahrara S, Campbell PL, Koch AE. CXCL16-mediated cell recruitment to rheumatoid arthritis synovial tissue and murine lymph nodes is dependent upon the MAPK pathway. *Arthritis Rheum.* 2006 Mar;54(3):765-78.

Sheikine Y, Hansson GK. Chemokines and atherosclerosis. Ann Med. 2004;36(2):98-118. Review. Sheikine Y, Hansson GK. Chemokines and atherosclerosis. *Ann Med.* 2004;36(2):98-118.

Sheikine Y, Bang CS, Nilsson L, Samnegard A, Hamsten A, Jonasson L, Eriksson P, Sirsjo A. Decreased plasma CXCL16/SR-PSOX concentration is associated with coronary artery disease. *Atherosclerosis.* 2006 Oct;188(2):462-6.

Shimaoka T, Kume N, Minami M, Hayashida K, Kataoka H, Kita T, Yonehara S. Molecular cloning of a novel scavenger receptor for oxidized low density lipoprotein, SR-PSOX, on macrophages. *J Biol Chem.* 2000 Dec 29;275(52):40663-6.

Shimaoka T, Nakayama T, Fukumoto N, Kume N, Takahashi S, Yamaguchi J, Minami M, Hayashida K, Kita T, Ohsumi J, Yoshie O, Yonehara S. Cell surface-anchored SR-PSOX/CXC chemokine ligand 16 mediates firm adhesion of CXC chemokine receptor 6-expressing cells. *J Leukoc Biol.* 2004 Feb;75(2):267-74.

Signoret N, Oldridge J, Pelchen-Matthews A, Klasse PJ, Tran T, Brass LF, Rosenkilde MM, Schwartz TW, Holmes W, Dallas W, Luther MA, Wells TN, Hoxie JA, Marsh M. Phorbol esters and SDF-1 induce rapid endocytosis and down modulation of the chemokine receptor CXCR4. *J Cell Biol.* 1997 Nov 3;139(3):651-64.

Springer TA. Traffic signals for lymphocyte recirculation and leukocyte emigration: The multistep paradigm. *Cell* 1994, Jan; 76(2):301-14.

Tabata S, Kadowaki N, Kitawaki T, Shimaoka T, Yonehara S, Yoshie O, Uchiyama T. Distribution and kinetics of SR-PSOX/CXCL16 and CXCR6 expression on human dendritic cell subsets and CD4+ T cells. *J Leukoc Biol.* 2005 May;77(5):777-86.

van der Voort R, van Lieshout AW, Toonen LW, Sloetjes AW, van den Berg WB, Figdor CG, Radstake TR, Adema GJ. Elevated CXCL16 expression by synovial macrophages recruits memory T cells into rheumatoid joints. *Arthritis Rheum.* 2005 May;52(5): 1381-91.

Wuttge DM, Zhou X, Sheikine Y, Wagsater D, Stemme V, Hedin U, Stemme S, Hansson GK, Sirsjo A. CXCL16/SR-PSOX is an interferon-gamma-regulated chemokine and scavenger receptor expressed in atherosclerotic lesions. *Arterioscler Thromb Vasc Biol.* 2004 Apr;24(4):750-5.

Wagsater D, Olofsson PS, Norgren L, Stenberg B, Sirsjo A. The chemokine and scavenger receptor CXCL16/SR-PSOX is expressed in human vascular smooth muscle cells and is induced by interferon gamma. *Biochem Biophys Res Commun.* 2004 Dec 24;325(4): 1187-93.

In: Chemokine Research Trends
Editor: Leah R. Grinwald, pp. 181-188

ISBN: 978-60021-588-9

Chapter 13

NOVEL ASPECTS OF AMNIOTIC FLUID EXAMINATION FOR CHORIOAMNIONITIS

Yoshikatsu Suzuki, Tamao Yamamoto, Ghada A. Daud, Nobuhiro Suzumori and Mitsuyo Tanemura
Department of Obstetrics and Gynecology,
Nagoya City University Medical Sciences, Kawasumi 1,
Mizuho-cho, Mizuho-ku, Nagoya, 467-8601, Japan

ABSTRACT

Chorioamnionitis, a major cause of pre-term labor, is generally diagnosed by histological examination of placenta tissues obtained from patients after delivery. Here we focus on use of amniotic fluid, which can be readily obtained in the clinical field during pregnancy, for diagnostic purposes. Culture is the gold standard for detection of infection, but bacteria often fail to be detected by routine culture in amniotic fluid obtained from patients complicated by pre-term labor, despite increase in infectious cytokines, such as interleukin-6 and interleukin-8. We have established detection of Escherichia coli (E coli), one major cause of chorioamnionitis, by a modified polymerase chain reaction technique.

There is considerable evidence that NO inhibits uterine contractility as well as cervical ripening during pregnancy, but in our study, the NO concentration was found to be increased in amniotic fluid obtained from patients complicated by chorioamnionitis. These results indicate that effects of NO may be attenuated for at least two reasons: the action of NO itself is down-regulated and the NO is rapidly degraded in the amniotic fluid.

Chorioamnionitis is a major cause of pre-term labor. As an alternative to histological examination of placenta tissues obtained from the patients after delivery it has been proposed that amniotic fluid be used for testing. For example, evaluation of inflammatory parameters, such as the presence of white blood cells [1.2.3], interleukin-6 [1.3-6], interleukin-8 [1.6.7] or leukocyte elastase [1.4.8.9] has been applied for diagnosis of intrauterine infection.

In the present report, we introduce new concepts for amniotic fluid obtained from patients complicated by chorioamnionitis.

ARE THERE BACTERIA IN AMNIOTIC FLUID?

Culture is the gold standard for detection of infection. However, in amniotic fluids obtained from patients complicated by chorioamnionitis, bacteria often fail to be detected by routine culture, even in cases with increase in various cytokines. There are two hypothesized explanations for the low detection rate of bacteria by culture in amniotic fluid: 1) limitations due to the presence of inhibitory substances in amniotic fluid, killing of bacteria by broad spectrum antibiotics, and small numbers of bacteria, 2) cytokines are activated not due to the existence of bacteria in amniotic fluid but are produced elsewhere.

The polymerase change reaction (PCR) for 16S ribosomal RNA (rRNA) has been porposed as am alternative to detect bacteria in amniotic fluid from patients with intrauterine infection [10]. However, this may show false positive results due to the high sensitivity of PCR. Furthermore, it has been found that E coli DNA may contaminate PCR reagents [11]. In our preliminary study, 16S rRNA primers could detect E coli DNA in amniotic fluid in half of the patients with chorioamnionitis, but the same rate was obtained for normal pregnant women, consistent with previous reports [10].

Populations of Escherichia coli (E coli), a major causative agent for intrauterine infection, are clonal and studies of the 72 strains of E coli of reference (ECOR) have allowed division into four main phylogenetic groups (A, A1, B2 and D) [12.13]. Strains of the ECOR B2 group appeared to be best suited to colonize the vagina and the prevalence of ECOR B2 group strains is high in amniotic fluid. Furthermore, pathogenicity of E coli is also correlated with the presence of virulence factors (VFs) encoded by eight genes in E coli strains responsible for extraintestinal infection; *neuC* (K1 capsule antigen), *hly* (alpha-hemolysin), *papC* (type P pilli), *sfa/foc-r* (type S pili and type 1C fimbriae), *fimH* (type 1 pili), *afa* (afimbrial adhesion), *iucC* (aerobactin), and *ibe A* (Ibe A protein). The *neuC*, *fimH*, *iucC* and *hly* are frequent in amniotic fluid [14]. Thus, both ECOR grouping and detection of VFs may be useful markers for pathogenic E coli.

We therefore investigated this possibility with amniotic fluid samples obtained from 18 patients complicated by chorioamnionitis (9 were aborted and 9 developed pre-term labor) and 21 normal pregnant women who underwent amniocentesis at early second trimester for detection of fetal chromosomal anomalies and delivered at term. The patients were diagnosed as having chorioamnionitis on histological examination of placental tissue (Blanc II or III) [6.13] after having aborted or delivered at preterm. In all cases, E coli were negative in standard culture. From amniotic fluid, DNA was extracted and used with forward and reverse primers for the ECOR grouping (*ChuA*, *yjaA* and *TSPE4.C2*) and the *VFs*, *fimH*, *iucC*, *papC*, *sfa/foc*, *hly*, *afa*, *new C* and *ibeA* [14] (Table1). Aliquots of DNA template were 2 µl and the PCR conditions were as follows; initial denaturation at 94 °C for 10 min followed by 32-35 cycles of denaturation at 94 °C for 15s, annealing at the melting temperature of 56 °C for 30s, extension at 72 °C for 1 min, followed by a final 10-minutes extension at 72 °C.

E coli bacteria were classified with the ECOR grouping; group B2 (*ChuA*+, *yjaA*+, TspE4.C2 +or-), group D (ChuA +, yjaA –, TspE4.C2 +or-), group B1 (ChuA –, yjaA +or-, TspE4.C2 +), group A (ChuA –, yjaA +or-, TspE4.C2 –). [12.13].

ATCC25922 (common E coli strain) at high concentration (more than 10^7 cells / ml) showed positive bands for *ChuA, YjaA* and *TspE4.C2*, and was classified as group B2. *YjaA* was negative at low concentrations (less than 10^6 cells/ml). *ChuA* also disappeared at 10^5

cells/ml (Fig.1A). With VFs, *fim H, sfa/foc, papC, iucC and ibeA* bands were seen in both groups at high concentrations (more than 10^7 cells / ml). However, they were not apparent at low concentration (Fig 1B).

Table 1. Nucleotide sequences of the various primers

Gene	Primer sequence	Tm(°C)	Size (bp)
ECOR grouping			
chuA-f	5'-GACGAACCAACGGTCAGGAT-3'	58	279
chuA-r	5'-TGCCGCCAGTACCAAAGACA-3'		
yjaA-f	5'-TGAAGTGTGTCAGGAGACGCTG-3'	58	211
yjaA-r	5'-ATGGAGAATGCGTTCCTCAAC-3'		
TSPE4.C2-f	5'-GAGTAATGTCGGGGCATTCA-3'	56	152
TSPE4.C2-r	5'-CGCGCCAACAAAGTACG-3'		
Virulence factors			
fimH-f	5'-TGCAGAACGGATAAGCCGTGG-3'	55	508
fimH-r	5'-GCAGTCACCTGCCCTCCGGTA-3'		
iucC-f	5'-AAACCTGGCTTACGCAACTGT-3'	55	269
iucC-r	5'-ACCCGTCTGCAAATCATGGAT-3'		
papC-f	5'-GACGGCTGTACTGCAGGGTGTGGCG-3'	65	328
papC-r	5'-ATATCCTTTCTGCAGGGATGCAATA-3'		
newC-f	5'-AGGTGAAAAGCCTGGTAGTGTG-3'	61	613
neuC-r	GGTGGTACATTCCGGGAGTC-3'		
sfa/foc-f	5'-GAGAACTGCTCCGGGTGCATACTCT-3'	65	410
sfa/foc-r	5'-CGGAGGAGTAATTACAAACCTGGCA-3'		
hly-f	5'-AACAAGGATAAGCACTGTTCTGGCT-3'	63	1177
hly-r	5'-ACCATATAAGCGGTCATTCCCGTCA-3'		
afa-f	5'-GCTGGGCAGCAAACTGATAACTCTC-3'	65	750
afa-r	5'-CATCAAGCTGTTTGTTCGTCCGCCG-3'		
ibeA-f	5'-TTACCGCCGTTGATGTTATCA-3'	60	171
ibeA-r	5'-CATTAGCTCTCGGTTCACGCT-3'		

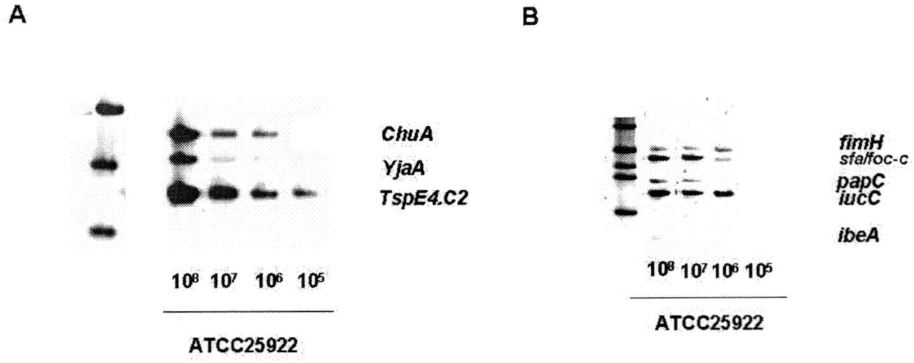

Figure 1. Evaluation of the ECOR group and VFs in common E coli strainsA. At 10^5-10^8 cell/ml of the common E coli strain ATCC25922, positive bands are evident at 279 bp for *ChuA*, 211 bp for *yja A* and 152 bp for *TSPE4.C4*, in ECOR grouping. B. At the same concentration of ATCC25922, positive bands are apparent at 508bp for *fimH*, 410 bp for *sfa/foc-c*, 328 bp for *papC*, 269 bp for *iucC* and 171 bp for *ibeA*.

In the 6 of 18 cases with chorioamnionitis, bands for *ChuA* and *Tspe4.C2* could be seen. Regarding VF primers, *neuC* was positive in 2/18, *fim H* in 3/18, *sfa/foc-c* in 3/18, *papC* in 1/18, *iucC* in 6/18 (two cases in normal pregnancy were positive) and *ibe A* in 1/18 in. Thus, in a total 18 cases with chorioamnionitis, E coli could be detected in 6 by ECOR grouping, and also in 6with reference to VFs. Four cases were positive with both (Table 2).

We conclude that both PCR ECOR grouping and PCR for VFs are useful to detect E coli in amniotic fluid and thus overcome difficulties with standard culture.

Does Increase in Nitric Oxide Regulate Uterine Contraction in Chorioamninitis?

Nitric oxide (NO) is well known to play crucial roles in female reproductive physiology [16-18]. Increase might be involved in normal hemodynamic changes occurring during pregnancy (maintenance of the uteroplacental circulation and control of maternal vascular tone) and there is considerable evidence that NO contributes to the regulation of uterine contractility as well as cervical ripening [18]. Reduction may contribute to the pathophysiology of intrauterine growth retardation [18] and preeclampsia [19, 20]. In our previous report, we described reduced action of NO to be probably due to down-regulation of cGMP, the second messenger of NO, in smooth muscle cells of resistant arteries in preeclampsia cases [19]. Severe infection, including septic shock, causes marked increase in NO via activation of inducible NO synthase accompanied with an increase in various cytokines in tissues affected by severe inflammation. In chorioamnionitis, increased concentrations of NO and interleukin-6 in amniotic fluid have also been reported [21.22], although how this might impact on the regulation of uterine contraction remains unclear.

Table 2. Details for the 18 patients complicated by chorioamninitis

					Amniotic	fluid		
Case	Delivery time (weeks)	Birth weight (g)	Blanc Classification of placenta	Presence of WBC	Interleukin-6 (pg/ml)	Leukocyte Erastase (μg/ml)	ECORgroup	VFs
1	23	520	2	-	6,680	12,400		
2	25	824	3	+	26,900	4,500		
3	28	1566	3	-	41,300	950	+	
4	24	463	3	+	784,000	3,000		+
5	26	886	3	+	69,600	12,100		
6	19	300	3	+	773,000	2,500		
7	19	278	3	+	114,000	18,000		
8	18	222	3	+	364,000	16,700	+	
9	26	745	2	+	70,000	12,000	+	+
10	16	198	2	+	125,000	2,200		
11	24	540	2	-	13,800	20,470		
12	20	348	2	+	10,900	8,000		
13	18	222	2	+	12,630	11,840		
14	16	88	3	+	12,480	11,780	+	+
15	29	1600	2	+	358,000	432,000	+	+
16	20	320	3	+	13,800	20,470		
17	24	708	2	-	10,900	18,900		
18	20	400	3	+	123,486	10,000	+	+

We therefore investigated possible roles of NO in cases of chorioamnionitis [23]. We assessed concentrations of NO, interleukin-6 and leucocyte elastase in amniotic fluid obtained from 8 cases complicated with chorioamnionitis diagnosed by histological examination of placenta (Blanc III) at 17-24 weeks of gestation, and compared with values for 89 pregnant women undergoing diagnostic amniocentesis in the early second trimester.

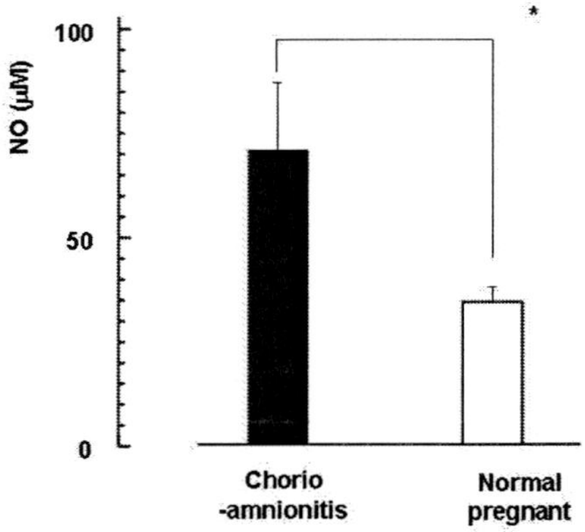

Figure 2. Levels of NO in amniotic fluid obtained from cases with chorioamnionitis and normal pregnant women. * $P<0.05$ *vs*. normal pregnant women.

The concentration of NO was significantly higher in the cases with chorioamnionitis than in the normal pregnant women (Fig. 2), along with the concentrations of interleukin-6 and leucocyte elastase. In the cases with chorioamnionitis, but not the normal pregnant women, the concentration of NO showed a tendency for an inverse correlation with leucocyte elastase ($R=0.755$, $P=0.07$).

Thus, we hypothesized that in cases with severe chorioamnionitis, despite increase NO production, the effects of NO may be attenuated for at least two reasons: the action of NO itself is down-regulated and the NO is rapidly degraded in the amniotic fluid.

REFERENCES

[1] Suzuki Y, Yamamoto T, Kojima K, Tanemura M, Tateyama H, Suzumori K. Evaluation levels of cytokines in amniotic fluid of women with intrauterine infection in the early second trimester. *Fetal Diag Ther* 2006; 21: 45-50.

[2] Romero R, Quintero R, Nores J, Avila C, Mazor M, Hanaoka S, Haqay Z, merchant L, Hobbins JC. Amniotic fluid white blood cell count: A rapid and simple test to diagnose microbial invasion of the amniotic cavity and predict preterm delivery. *Am J Obstet Gynecol*. 1991; 165: 821-830.

[3] Romero R, Yoon BH, Mazor M, Gomez R, Gonzalez R, Diamond MP, Baumann P, Araneda H, Kenny JS, Cotton DB, Shegal P. A comparative study of the diagnostic performance of amniotic fluid glucose, white blood cell count, interleukin-6, and Gram stain in the detection of microbial invasion in patients with preterm premature rupture of membranes. *Am J Obstet Gynecol* 1993; 169: 839-851.

[4] Romero R, Yoon BH, Kenney JS, Gomez R, Allison AC, Sehgal PB. Amniotic fluid interleukin-6 determinations are of diagnostic and prognostic value in preterm labor. *Am J Reprod Immunol* 1993; 30: 167-183.

[5] Tsuda A, Ikegami T, Hirano H, Sanada H, Ogawa M, Sasaki M, Tanaka T. The relationship between amniotic fluid interleukin-6 concentration and histologic evidence of chorioamnionitis. *Acta Obstet Gynecol Scand* 1998; 77: 515-520.

[6] Suzuki Y, Yamamoto T, Kojima K, Tanemura M, Tateyama H, Suzumori K. Evaluation of levels of cytokines in amniotic fluid of women with intrauterine infection in the early second trimester. *Fetal Diagn Ther* 2006; 21: 45-50.

[7] Romero R, Ceska M, Avila C, Mazor M, Behnke E, Linedley I. Neutrophil attractant/activating peptide-1/interleukin-8 in term and preterm parturition. *Am J Obstet Gynecol.* 1991; 165: 813-820.

[8] Romero R, Emamian M, Wan M, et al. The value of the leukocyte esterase test in diagnosing intra-amniotic infection. *Am J Perinatol.* 1988; 5: 64-69.

[9] Hoskins IA, Johnson TRB, Winkel CA. Leukocyte esterase activity in human amniotic fluid for the rapid detection of chorioamnionitis. *Am J Obstet Gynecol.* 1987; 157: 730-732.

[10] Markenson GR, Martin RK, Tillotson-Criss M, Foley KS, Stewart RS Jr, Yancey M. The use of the polymerase chain reaction to detect bacteria in amniotic fluid in pregnancies complicated by preterm labor. *Am J Obstet Gynecol* 1997; 177: 1471-1477.

[11] Meier A, Persing DH, Finken M, Boettger EC. Elimination of contaminating DNA within polymerase chain reaction reagents: implications for a general approach to detection of uncultured pathogens. *J Clin Microbiol* 1993; 31: 646-652.

[12] Desjardin PB, Picard BK, Elion J, Denamur E. Sex in Esherichia coli does not disrupt the clonal structure of the population: evidence from random amplified polymorphic DNA and restriction-fragment-length polymorphism. *J Mol Evol* 1995; 41: 440-448.

[13] Herzer PJ, Inouye S, Inouye M, Whittam TS. Phylogenetic distribution of branched RNA-linked multiscopy single-stranded DNA among natural isolates of Escherichia coli. *J bacterial* 1991; 172: 6175-6181.

[14] Watt S, Lanotte P, Mereghetti L, Moulin-Schouleur M, Picard B, Quentin R. Escherichia coli strains from pregnant women and neonates: Intraspecies genetic distribution and prevalence of virlulence factors. *J Clin Microbiol* 2003; 41: 1929-1935.

[15] Blanc WA. Pathology of the placenta and cord in ascending and in haematogenous infection. *Ciba Found Symp* 1979; 77: 17-38.

[16] Chwalisz K, Bushimischi I, Garfield RE. Role of nitric oxide in obstetrics. *Prenat Neonat Med* 1996; 1: 292-329.

[17] Rosseli M. Nitric oxide and production. *Mol Hum Reprod* 1997; 3: 639-641.

[18] Ledingham MA, Thomson AJ, Greer IA, Norman JE. Nitric oxide in parturition. *Br J Obstet Gynaecol* 2000; 107: 581-593.

[19] Suzuki Y, Kajikuri J, Suzumori K, Itoh T. Mechanisms underlying the reduced endothelium-dependent relaxation in human omental resistance arteries in pre-eclampsia. *J Physiol* 2000; 527: 163-174.

[20] Yamamoto T, Suzuki Y, Kojima K, Suzumori K. Reduced flow-mediated vasodilatation is not due to a decrease in production of nitric oxide in preeclampsia, *Am J Obstet Gynecol* 2005: 192: 558-563.

[21] Hus CD, Aversa K, Meaddough E, Lu LC, Copel JA. Elevated amniotic fluid nitric oxide metabolites and cyclic guanosine 3', 5'-monophosphate in pregnant women with intra-amniotic infection. *Am J Obstet Gynecol* 1997; 177: 793-796.
[22] Hus CD, Meaddough E, Hong SF, Aversa K, Lu LC, Copel JA. Elevated amniotic fluid nitric oxide metabolites and interleukin-6 in intra-amniotic infection. *J Soc Gynecol Invest* 1998; 5: 21-24.
[23] Ghada DA, Suzuki Y, Yamamoto T, Suzumori K, Togari H. Reduced nitric oxide in amniotic fluid of patients with chorioamnionitis. *Fetal Diagn Ther* 2006; 21: 77- 80.

In: Chemokine Research Trends
Editor: Leah R. Grinwald, pp. 189-193

ISBN: 978-60021-588-9

Chapter 14

ANTI-INFLAMMATORY EFFECTS OF NIFEDIPINE, A DIHYDROPYRIDINE-BASED CALCIUM BLOCKER, ON ENDOTHELIAL CELLS

Sho-ichi Yamagishi*[*], *Kazuo Nakamura and Takanori Matsui
Department of Medicine, Division of Cardiovascular Meedicine, Kurume University School of Medicine, Kurume 830-0011, Japan

ABSTRACT

Endothelial cell (EC) dysfunction has been proposed to be an initial event that leads to the development of atherosclerosis. There is a growing body of evidence that atherosclerosis is an inflammatory disease, and ECs are target for chemokines and growth factors released during inflammation. Recently, nifedipine, one of the most widely used dihydropyridine-based calcium antagonists (DHPs) for the treatment of patients with hypertension, has been shown to inhibit vascular inflammation and subsequently improve endothelial dysfunction in many cardiovascular diseases, thus slowing the development and progression of atherosclerosis. However, the molecular mechanisms underlying this are not fully understood, because ECs do not possess voltage-operated L-type calcium channels. In this review, we discuss molecular mechanisms by which nifedipine could exert anti-inflammatory effects on ECs.

Key Words; cardiovascular disease, chemokines, nifedipine, oxidative stress

INTRODUCTION

Endothelial cell (EC) dysfunction has been proposed to be an initial event that leads to the development of atherosclerosis [1,2]. Many lines of evidence show that atherosclerosis is an inflammatory disease, and ECs are target for chemokines and growth factors released during the inflammation [2].

[*] E-mail address: shoichi@med.kurume-u.ac.jp Tel: +81-942-31-7580 Fax: +81-942-31-7707

Recently, nifedipine, one of the most widely used dihydropyridine-based calcium antagonists (DHPs) for the treatment of patients with angina and hypertension [3,4], has been shown to improve endothelial dysfunction in patients with coronary risk factors [5-9]. Since endothelial dysfunction is one of initial steps of atherosclerosis [10], nifedipine could slow the development and progression of atherosclerosis by restoring endothelial dysfunction [11,12]. Furthermore, this anti-atherosclerotic activity may be, at least in part, independent of the blood pressure lowering effect [13]. However, the molecular mechanisms underlying this are not fully understood, because ECs do not possess voltage-operated L-type calcium channels, which are thought to be a target for DHPs [14]. In this review, we discuss molecular mechanisms by which nifedipine could exert anti-inflammatory effects on ECs.

ANTI-INFLAMMATORY AND ANTI-OXIDATIVE PROPERTIES OF NIFEDIPINE

There is a growing body of evidence, ranging from *in vitro* experiments to pathologic analysis to epidemiologic studies, that atherosclerosis is intrinsically an inflammatory disease [15]. The classic pro-inflammatory cytokines such as tumor necrosis factor-α (TNF-α) and interleukin-1 induce pro-thrombotic-pro-inflammatory reactions in ECs, thus being initially involved in the pathogenesis of atherosclerosis. Upon exposure to these cytokines, ECs have been reported to undergo profound functional changes that include expression of adhesion molecules, production of monocyte chemoattractant protein (MCP-1), and induction of pro-coagulant activity [16-18]. It is well known that a number of pathophysiological stimuli induce EC activation as a function of NADPH oxidase-mediated ROS-induced signal transduction [19-23].

One early phase of atherosclerosis involves the recruitment and firm adhesion of inflammatory cells to ECs [15]. This process is predominantly mediated by MCP-1 and vascular cell adhesion molecule-1 (VCAM-1) expression in ECs. Indeed, the selective targeting of MCP-1 or VCAM-1 was shown to markedly decrease atheromataous lesion formation in apolipoprotein E knockout mice [24,25]. We have recently found that 1 μ M nifedipine inhibits the TNF-α-induced reactive oxygen species (ROS) generation and subsequent MCP-1 and VCAM-1 expression in human cultured ECs by suppressing NADPH oxidase activity [26,27]. The plasma concentration of nifedipine after the oral administration of a single preparation of 40-mg tablet reaches a maximum at 3 h, and the peak value is reported to be about 0.2 μM [28]. However, localized concentrations of nifedipine in ECs are likely to be above 0.2 μM because of its high affinity to plasma membranes [28]. So, the concentration of nifedipine having anti-inflammatory effects on ECs (1 μM) may be comparable to the therapeutic levels, which are achieved in the treatments of patients with hypertension [30]. Taken together, these observations suggest that nifedipine could play a protective role against atherosclerosis via inhibition of vascular inflammation, an initial step of atherosclerosis.

The exact molecular mechanism underlying the anti-oxidative properties of nifedipine on ECs remains to be elucidated because ECs do not possess voltage-operated L-type calcium channels, which were thought to be a target for DHPs [14]. However, nifedipine is a lipid soluble DHP with high affinity to plasma membranes [29]. Therefore, it is conceivable that this compound may be incorporated into cell surface membrane and subsequently inhibit the TNF-α-signaling pathways to NADPH oxidase activation in ECs. The redox-sensitive transcription factor NF-κB modulates gene expression of pro-inflammatory cytokines and growth factors, thus playing a central role in the pathogenesis of vascular dysfunction and atherosclerosis as well [31]. Since TNF-α up-regulates MCP-1 and VCAM-1 mRNA levels through transcriptional activation of NF-κB [32,33], it is also probable that nifedipine could have decreased the MCP-1 and VCAM-1 expression in ECs by inhibiting ROS generation and subsequent NF-κB activation. Nifedipine also inhibits the activation of NF-κB in cultured fibroblasts, further supporting our speculation [34].

In addition, we have recently found that nifedipine also inhibits the TNF-α–induced up-regulation of CD40 mRNA levels in ECs. Furthermore, antibody raised against human CD40 significantly blocks the TNF-α–induced up-regulation of MCP-1 mRNA levels in ECs. These findings suggest that nifedipine could inhibit the TNF-α-induced up-regulation of MCP-1 mRNA levels via suppression of CD40, a cell surface receptor that belongs to the TNF-α receptor, in ECs [35].

Perspectives

We reviewed here the molecular mechanisms by which nifedipine could exert anti-inflammatory effects on ECs. The clinical relevance of this DHP should be further studied in hypertensive patients associated with the metabolic syndrome or diabetes.

References

[1] Ross R. Atherosclerosis: an inflammatory disease. *N Eng J Med* 1999; 340: 115-26.

[2] Introna M, Mantovani A. Early activation signals in endothelial cells. *Arterioscler Thromb Vasc Biol* 1997; 17: 423-8.

[3] Yamagishi S, Nakamura K, Takenaka K, Matsui T, Inoue H. Pleiotropic effects of nifedipine on atherosclerosis. *Curr Pharm Des* 2006; 12: 1543-7.

[4] Brown MJ, Palmer CR, Castaigne A, de Leeuw PW, Mancia G, et al. Morbidity and mortality in patients randomised to double-blind treatment with a long-acting calcium-channel blocker or diuretic in the International Nifedipine GITS study: Intervention as a goal in hypertension treatment (INSIGHT). *Lancet* 2000; 356: 366-72.

[5] Taddei S, Virdis A, Ghiadoni L, Sudanom I, Salvetti A. Antihypertensive drugs and reversing of endothelial dysfunction in hypertension. *Curr Hypertens Rep* 2000; 2: 64-70.Verhaar MC, Honing ML, van Dam T, Zwart M, Koomans HA, et al. Nifedipine improves endothelial function in hypercholesterolemia, independently of an effect on blood pressure or plasma lipids. *Cardiovasc Res* 1999; 42: 752-.

[6] ENCORE Investigators. Effect of nifedipine and cerivastatin on coronary endothelial function in patients with coronary artery disease: the ENCORE I Study (Evaluation of Nifedipine and Cerivastatin On Recovery of coronary Endothelial function). *Circulation* 2003; 10: 422-8.

[7] Kitakaze M, Asanuma H, Takashima S, Minamino T, Ueda Y, Sakata Y, Asakura M, Sanada S, Kuzuya T, Hori M. Nifedipine-induced coronary vasodilation in ischemic hearts is attributable to bradykinin- and NO-dependent mechanisms in dogs. *Circulation* 2000; 101: 311-7.

[8] Yamasaki K, Aoki M, Makino H, Hashiya N, Shimizu H, Ohishi M, Ogihara T, Morishita R. Effects of nifedipine on endothelial function in normotensive smokers: potential contribution of increase in circulating hepatocyte growth factor. J Human Hypertens 2004; 18: 701-5.

[9] Zeiher AM, Schachinger V. Coronary endothelial vasodilator dysfunction: clinical relevance and therapeutic implications. *Z Kardiol* 1994; 83: S7-14.

[10] Simon A, Gariepy J, Moyse D, Levenson J. Differential effects of nifedipine and co-amilozide on the progression of early carotid wall changes. *Circulation* 2001; 103: 2949-54.

[11] Lichtlen PR, Hugenholtz PG, Rafflenbeul W, Hecker H, Jost S, Deckers JW. Retardation of angiographic progression of coronary artery disease by nifedipine. Results of the International Nifedipine Trial on Antiatheroslerotic Therapy (INTACT). INTACT Group Investigator. *Lancet* 1990; 335: 1109-13.

[12] Nayler WG. Review of preclinical data of calcium channel blockers and atherosclerosis. *J Cardiolvasc Pharmacol* 1999; 33: S7-11.

[13] Mark IT, Weglicki WB. Antioxidant activity of calcium channel blocking drugs. *Methods Enzymol* 1994; 234: 620-30.

[14] Manitovani A, Bussolino F, Dejana E. Cytokine regulation of endothelial cell function. *FASEB J* 1992; 6: 2591-9.

[15] Vanden Berghe W, Vermeulen L, De Wilde G, De Bosscher K, Boone E, Haegeman G. Signal transduction by tumor necrosis factor and gene regulation of the inflammatory cytokine interleukin-6. *Biochem Pharmacol* 2000; 60: 1185-95.

[16] Mackay F, Loetscher H, Stueber D, Gehr G, Lesslauer W. Tumor necrosis factor-alpha (TNF-alpha)-induced cell adhesion to human endothelial cells is under dominant control of one TNF receptor type. *J Exp Med* 1992; 177: 1277-86.

[17] Nawroth PP, Stern DM. Modulation of endothelial cells hemostatic properties by tumor necrosis factor. *J Exp Med* 1986; 163: 740-5.

[18] Chen XL, Zhang Q, Zhao R, Medford RM. Superoxide, H2O2, and iron are required for TNF-alpha-induced MCP-1 gene expression in endothelial cells: role of Rac1 and NADPH oxidase. Am J Physiol Heart *Circ Physiol* 2004; 286: H1001-7.

[19] Frey RS, Rahman A, Kefe, JC, Minshall RD, Malik AB. PKCzeta regulates TNF-alpha-induced activation of NADPH oxidase in endothelial cells. *Circ Res* 2002; 90: 1012-9.

[20] Zhang WJ, Frei B. Alpha-lipoic acid inhibits TNF-α-induced NF-kappaB activation and adhesion molecule expression in human aortic endothelial cells. *FASEB J* 2001; 15: 2423-32.

[21] Griendling KK, Sorescu D, Ushio-Fukai M. NAD(P)H oxidase: role in cardiovascular biology and disease. *Cir Res* 2000; 86: 494-501.

[22] Kunsch C, Medford RM. Oxidative stress as a regulator of gene expression in the vasculature. *Cir Res* 1999; 85: 753-66.

[23] Gu L, Okada Y, Clinton SK, Gerard C, Sukhova GK, et al. Absence of monocyte chemoattractant protein-1 reduces atherosclerosis in low density lipoprotein receptor-deficient mice. *Mol Cell* 1998; 2: 275-81.

[24] Dansky HM, Barlow CB, Lominska C, Sikes JL, Kao C, et al. Adhesion of monocytes to arterial endothelium and initiation of atherosclerosis are critically dependent on vascular cell adhesion molecule-1 gene dosage. *Arterioscler Thromb Vasc Biol* 2001; 21: 1662-7.

[25] Yamagishi S, Inagaki Y, Kikuchi S. Nifedipine inhibits tumor necrosis factor-alpha-induced monocyte chemoattractant protein-1 overexpression by blocking NADPH oxidase-mediated reactive oxygen species generation. *Drugs Exp Clin Res* 2003; 29: 147-52.

[26] Yamagishi S, Takeuchi M. Nifedipine inhibits tumor necrosis factor-α-induced leukocyte adhesion to endothelial cells by suppressing vascular cell adhesion molecule-1 (VCAM-1) expression. *Drugs Exp Clin Res* 2004; 30: 163-8.

[27] Ramsch KD, Ziegler R. Plasma concentrations of various nifedipine formulations in healthy volunteers. 6th International Adalat *Symposium. Excepta Medica* 1986; 23-32.

[28] Herbette LG, Vant Erve YM, Rhodes DG. Interaction of 1,4 dihydropyridine calcium channel antagonists with biological membranes: lipid bilayer partitioning could occur before drug binding to receptors. *J Mol Cell Cardiol* 1989; 21: 187-201.

[29] Fukuo K, Yang J, Yasuda O, Mogi m, Suhara T, et al. Nifedipine indirectly upregulates superoxide dismutase expression in endothelial cells via vascular smooth muscle cell-dependent pathways. *Circulation* 2002; 106: 356-61.

[30] Barnes PJ, Karin M. Nuclear factor-kappaB: a pivotal transcription factor in chronic inflammatory diseases. *N Engl J Med* 1997; 336: 1066-71.

[31] Ciesielski CJ, Andreakos E, Foxwell BM, Feldmann M. TNFalpha-induced macrophage chemokine secretion is more dependent on NF-kappaB expression than lipopolysaccharides-induced macrophage chemokine secretion. *Eur J Immunol* 2002; 32: 2037-45.

[32] Chen YH, Lin SJ, Chen JW, Ku HH, Chen YL. Magnolol attenuates VCAM-1 expression in vitro in TNF-alpha-treated human aortic endothelial cells and in vivo in the aorta of cholesterol-fed rabbits. *Br J Pharmacol* 2002; 135: 37-47.

[33] Matsumori A, Nunokawa Y, Sasayama S. Nifedipine inhibits activation of transcription factor NF-kappaB. *Life Sci* 2000; 67 2655-61.

[34] Yamagishi S, Takeuchi M. Nifedipine inhibits tumor necrosis factor-α-induced up-regulation of monocyte chemoattractant protein-1 mRNA levels by suppressing CD40 expression in endothelial cells. *Drugs Exp Clin Res* 2005; 31: 13-7.

INDEX

A

D

E

F

G

H

I

J

K

L

M

N

O

P

R

S

T

U

V

W

Y